T0258715

Social Touch in Human–Robot Interaction

In this book for researchers and students, editors Shiomi and Sumioka bring together contributions from researchers working on the CREST project at ATR Deep Interaction Laboratories, a world leader in social robotics, to comprehensively describe robot touch systems from hardware to applications.

Appropriate touch from robots to humans is essential for social robots, but achieving this requires various solutions at every stage of the touch process. Through this book, readers will gain an understanding of the needs, essential systems and communication cues, behaviour designs, and real-world issues for social touch applications. This book compiles and updates technical and empirical research that was previously scattered throughout the literature into a single volume. Through individually authored chapters addressing various elements of ATR's CREST project, this book tackles key areas where understanding is needed to realize acceptable touch interaction, including pre-touch interaction, interaction design for touching and being touched, behaviour changes caused by touch interaction, and applications of social touch interaction. It introduces a touch sensor and robots developed by the authors, including several touch-related behaviours and design policies. This approach will enable readers to easily apply this knowledge to their own social robotics programs. This book is invaluable for anyone who wishes to understand and develop social robots that physically interact with people.

It is most beneficial for researchers and upper undergraduate and graduate students in the fields of human–robot/agent/computer interaction and social touch interaction and those in the broader fields of engineering, computer science, and cognitive science.

Masahiro Shiomi is a senior research scientist and a group leader of the Agent Interaction Design Laboratory at the Interaction Science Laboratories at ATR, Kyoto, Japan. He is also a visiting professor at Kobe University. He earned an MEng and a PhD in engineering from Osaka University in 2004 and 2007, respectively. He served as a program chair on Ro-MAN2020, HAI2021, and HRI2022 and as an associate editor for the *ACM Transactions on Human–Robot Interaction.*

Hidenobu Sumioka is a group leader of Presence Media Research Group in Hiroshi Ishiguro Laboratories, ATR, Kyoto, Japan. He is also a visiting associate professor at Kobe University. He earned a PhD in engineering from Osaka University in 2008.

Social Touch in Human–Robot Interaction
Symbiotic Touch Interaction between Human and Robot

Edited by
Masahiro Shiomi
and Hidenobu Sumioka

CRC Press
Taylor & Francis Group
Boca Raton London New York

CRC Press is an imprint of the
Taylor & Francis Group, an **informa** business

Designed cover image: 3D represents the research of artificial intelligence AI to develop robots and cyborgs for future people to live. Digital data mining and machine learning technology design for computer brains.

First edition published 2024
by CRC Press
2385 NW Executive Center Drive, Suite 320, Boca Raton FL 33431

and by CRC Press
4 Park Square, Milton Park, Abingdon, Oxon, OX14 4RN

CRC Press is an imprint of Taylor & Francis Group, LLC

ISBN: 978-1-032-47026-9 (hbk)
ISBN: 978-1-032-47025-2 (pbk)
ISBN: 978-1-003-38427-4 (ebk)

DOI: 10.1201/9781003384274

Typeset in Minion
by codeMantra

Contents

Contributors

Qi An is an associate professor at the University of Tokyo and a visiting researcher at RIKEN. He received his PhD in engineering from the University of Tokyo.

Emi Anzai is a lecturer at Nara Women's University, Nara, Japan. She received her PhD from Ochanomizu University in 2017.

Tomo Funayama is a research engineer at Hiroshi Ishiguro Laboratories, ATR, Kyoto, Japan. He received his master's degree from Osaka University in 2017.

Norihiro Hagita is the chair and professor in the Art Science Department at Osaka University of Arts, and the director and ATR fellow of Norihiro Hagita Laboratories at Advanced Telecommunications Research Institute International (ATR).

Koki Haruno was a student at the Department of Electrical, Electronic and Mechanical Engineering, Graduate School and Faculty of Engineering, Osaka Institute of Technology. He graduated in March 2023.

Ayumi Hayashi was an intern student at Interaction Science Laboratories, ATR, Kyoto, Japan. She received her bachelor's degree from Nara Women's University in 2021.

Haruhiro Higashida was a specially appointed professor at the Research Center for Child Mental Development, Kanazawa University, until March 2023.

Takahiro Hirano was an intern student at Interaction Science Laboratories, ATR, Kyoto, Japan. He received his master's degree from Doshisha University in 2018.

Hiroshi Ishiguro is a distinguished professor at the Graduate School of Engineering Science, Osaka University. He received his PhD in systems engineering from Osaka University, Japan, in 1991.

Nobuhiro Jinnai was a student at the Graduate School of Engineering Science, Osaka University. He graduated in March 2018.

Higashino Kana was an intern student at the Interaction Science Laboratories, ATR, Kyoto, Japan. She received her master's degree from Doshisha University in 2022.

Masayuki Kanbara is an associate professor at Nara Institute of Science and Technology, Nara, Japan. He received his PhD from Nara Institute of Science and Technology in 2002.

Mitsuhiko Kimoto is a researcher at the Interaction Science Laboratories at ATR, Kyoto, Japan. He received his PhD from Doshisha University in 2019.

Youji Kohda is a professor in the Graduate School of Knowledge Science at the Japan Advanced Institute of Science and Technology (JAIST), Japan. He received his PhD from the University of Tokyo.

Atsumu Kubota was an intern student at the Interaction Science Laboratories, ATR, Kyoto, Japan. He received his master's degree from Doshisha University in 2022.

Hirokazu Kumazaki is currently a professor at Nagasaki University, School of Medicine. He received his medical degree in 2015 from Keio University, Japan.

Ryo Kurazume is a professor at the Graduate School of Information Science and Electrical Engineering, Kyushu University. He received his PhD in mechanical engineering from Tokyo Institute of Technology.

Takamasa Iio is an associate professor at Doshisha University, Kyoto, Japan. He received his PhD from Doshisha University in 2012.

Dario Alfonso Cuello Mejía is a researcher at the Interaction Science Laboratories at ATR, Kyoto, Japan. He received his PhD from Osaka University in 2023.

Masaru Mimura was a professor in the Department of Neuropsychiatry, Keio University School of Medicine, Tokyo, Japan, until March 2023. Now, he is a specially appointed professor at Keio University School of Medicine.

Takashi Minato is a team reader for the Interactive Robot Research Team at RIKEN Information R&D and Strategy Headquarters. He received his PhD from Osaka University.

Taro Muramatsu is an associate professor at Keio University, Japan. He is one of the few experts in judicial psychiatry in Japan.

Kohei Nakajima is an associate professor at the University of Tokyo. He received his PhD from the University of Tokyo.

Junya Nakanishi is an assistant professor at Osaka University. He received his PhD from Osaka University.

Aya Nakata was an intern student at the Interaction Science Laboratories, ATR, Kyoto, Japan. She received her master's degree from Nara Institute of Science and Technology in 2017.

Hiroshi Nittono is a professor at Osaka University, Osaka, Japan. He received his PhD from Osaka University in 1998.

Yuka Okada was an intern student at the Interaction Science Laboratories, ATR, Kyoto, Japan. She received her master's degree from Doshisha University in 2022.

Aoba Saito was an intern student at the Interaction Science Laboratories, ATR, Kyoto, Japan. He received his master's degree from Doshisha University in 2020.

Naoki Saiwaki is a professor at Nara Women's University, Nara, Japan. He received his PhD from Osaka University in 1993.

Kurima Sakai is a researcher at Hiroshi Ishiguro Laboratories, ATR, Kyoto, Japan. He received his PhD in engineering from Osaka University in 2017.

Kodai Shatani was an intern student at Hiroshi Ishiguro Laboratories, ATR, Kyoto, Japan. He received his master's degree from Osaka University in 2018.

Katsunori Shimohara was a professor in the Department of Information Systems Design, Faculty of Science and Engineering in the Graduate School of Science and Engineering, Doshisha University, Kyoto, Japan. He received his PhD from Kyushu University in 2000.

Masahiro Shiomi is a senior research scientist and a group leader of the Agent Interaction Design Laboratory at the Interaction Science Laboratories at ATR, Kyoto, Japan. He is also a visiting professor at Kobe University. He received his PhD in engineering from Osaka University in 2007.

Hidenobu Sumioka is a group leader of Presence Media Research Group in Hiroshi Ishiguro Laboratories, ATR, Kyoto, Japan. He is also a visiting associate professor at Kobe University. He received his PhD in engineering from Osaka University in 2008.

Takashi Takuma is a professor at the Department of Electrical and Electronic Systems Engineering, Faculty of Engineering, Osaka Institute of Technology. He received his PhD from Osaka University.

Nobuo Yamato is a student at the Graduate School of Knowledge Science at the Japan Advanced Institute of Science and Technology (JAIST), Japan.

Yuichiro Yoshikawa is an associate professor at the Graduate School of Engineering Science, Osaka University. He received his PhD in engineering from Osaka University, Japan, 2005.

Teruko Yuhi is a technical staff at the Research Center for Child Mental Development, Kanazawa University, Japan.

Xiqian Zheng was an intern student at Hiroshi Ishiguro Laboratories, ATR, Kyoto, Japan. He received his PhD from Osaka University in 2022. Please add an author info.

SECTION 1

Introduction

CHAPTER 1

Introduction to Social Touch in Human–Robot Interaction

Masahiro Shiomi and Hidenobu Sumioka

Advanced Telecommunications Research
Institute International, Kyoto, Japan

1.1 SOCIAL TOUCH IN HUMAN–HUMAN INTERACTION

Touch, which is an indispensable element of human–human interaction, conveys much more than affection, concern, or care; it is a medium for expressing a myriad of intentions and emotions. When analyzed from a bottom-up perspective, the relationship between low-threshold unmyelinated peripheral afferent fibers (also known as C-touch or CT fibers) and the characteristics of naturalistic affiliative interpersonal touch becomes critical [1]. On the other hand, viewing social touch from a top-down perspective emphasizes that it is predicated on interaction within interpersonal relationships, particularly reciprocal ones between humans [2].

Beyond the realm of personal interactions, social touch plays an essential role in wider developmental and emotional contexts. It is important in child development and the enhancement of parent–child relationships [3]. In terms of mental health, touch serves as a therapeutic tool that contributes to the emotional well-being of patients with Autism Spectrum Disorder (ASD) [4]. Social touch remains an influential factor that shapes interpersonal relationships and contributes to the overall psychological well-being of adults [5]. Even though social touch might not be literally required for

DOI: 10.1201/9781003384274-2

3

survival, it obviously plays a critical role in scores of life aspects, from early development stages to maintaining well-being in adulthood.

Researchers have investigated social touch in human–human interaction because touch behaviors serve as a channel for non-verbal communication such as gaze and gestures and express wider social signals [2]. Moreover, cultural background significantly influences touch behaviors as social signals in interaction. Similar to language, different societies adhere to different social norms, even in touch interaction [6]. Therefore, understanding such cultural differences is also essential in social touch studies in human–human interaction.

1.2 SOCIAL TOUCH IN HUMAN–ROBOT INTERACTION

With their physical bodies, robots can physically interact with humans just as we humans do with each other. In fact, human–robot touch interactions provide various benefits for human beings, such as the advantages of social touch among humans [7,8]. These studies showed a new aspect of social robots; many past human–robot interaction studies focus on conversational interaction, although such a possibility enables using social robots as physically interactive partners for human beings.

What defines a touch as *social* when human–robot interaction is involved? We believe that social touch involves recognizing a relationship with a robot, much like relationships with other people. Being hit by a ball is merely physical contact. However, if an entity that seems to possess human-like features and intent touches you, it's perceived as a *social touch*. This idea isn't limited to positive relationships: even in adversarial situations, the same principle applies. Your pet robot is quite welcome to touch you; being touched by a disliked robot will undoubtedly intensify that negative feeling. A social touch with robots resembles an amplifier in relationships.

To understand social touch in human–robot interaction, the design and implementation of appropriate hardware (e.g., robots and sensors) and software (e.g., a framework of social touch interaction) is crucial. Social robots that physically interact with people need abilities that initiate, respond to, and modulate social touch interaction based on the interaction contexts and relationships with interacting people. Designing such anthropomorphic robots and their attributions of intentionality is also essential for acceptable social touch interaction with people. In addition, evaluating the effects of social touch applications by robots is crucial for understanding how their social touch interactions influence people's decision-making and behaviors and support their daily lives.

In the context of social touch between humans and robots, robots must learn the norms of touch. As machine learning evolves, we anticipate robots that interact with a smoother, more natural touch, potentially replicating familiar gestures. However, such a computational approach relies on understanding the standard patterns of touch in advance. For this purpose, we are establishing computational social touch, a computational theory that calculates and reproduces these interactions at the algorithmic level to achieve infrastructure for *symbiotic interaction between humans and robots* through social touch. We are currently analyzing human touch behaviors and exploring how robots mimic such movements to advance the field of computational social touch.

This book presents an in-depth examination of social touch in the context of human–robot interaction, investigating the necessary technologies and emerging applications in this field. Our objective is to provide a comprehensive overview of this multidisciplinary domain, which converges at the intersection of robotics, social sciences, and engineering. This book is intended for researchers and students working in the field of human–robot interaction.

REFERENCES

[1] F. McGlone, A. B. Vallbo, H. Olausson, L. Loken, and J. Wessberg, "Discriminative touch and emotional touch," *Canadian Journal of Experimental Psychology/Revue canadienne de psychologie expérimentale*, vol. 61, no. 3, pp. 173, 2007.

[2] A. Gallace, and C. Spence, "The science of interpersonal touch: An overview," *Neuroscience & Biobehavioral Reviews*, vol. 34, no. 2, pp. 246–259, 2010.

[3] C. J. Cascio, D. Moore, and F. McGlone, "Social touch and human development," *Developmental Cognitive Neuroscience*, vol. 35, pp. 5–11, 2019.

[4] M. Coeckelbergh, C. Pop, R. Simut, A. Peca, S. Pintea, D. David, and B. Vanderborght, "A survey of expectations about the role of robots in robot-assisted therapy for children with ASD: Ethical acceptability, trust, sociability, appearance, and attachment," *Science and Engineering Ethics*, vol. 22, no. 1, pp. 47–65, 2015.

[5] T. Field, "Touch for socioemotional and physical well-being: A review," *Developmental Review*, vol. 30, no. 4, pp. 367–383, 2010.

[6] M. S. Remland, T. S. Jones, and H. Brinkman, "Interpersonal distance, body orientation, and touch: Effects of culture, gender, and age," *The Journal of Social Psychology*, vol. 135, no. 3, pp. 281–297, 1995.

[7] C. J. Willemse, and J. B. van Erp, "Social touch in human-robot interaction: Robot-initiated touches can induce positive responses without extensive prior bonding," *International Journal of Social Robotics*, vol. 11, no. 2, pp. 285–304, 2019.

[8] M. Shiomi, H. Sumioka, and H. Ishiguro, "Survey of social touch interaction between humans and robots," *Journal of Robotics and Mechatronics*, vol. 32, no. 1, pp. 128–135, 2020.

SECTION 2

System Development for Social Touch Interaction

Development of Fabric Sensor System toward Natural Pre-touch and Touch Interaction

Hidenobu Sumioka

Advanced Telecommunications Research Institute International, Kyoto, Japan

Takashi Minato

RIKEN, Saitama, Japan

Qi An

The University of Tokyo, Tokyo, Japan

Ryo Kurazume

Kyushu University, Fukuoka, Japan

Masahiro Shiomi

Advanced Telecommunications Research Institute International, Kyoto, Japan

DOI: 10.1201/9781003384274-4

2.1 INTRODUCTION

Despite the remarkable advances in the development of social robots that are designed to engage with humans in everyday settings, the interaction modalities between such robots and humans remain constrained, particularly in the realm of tactile engagement. While most social robots are equipped to communicate through gestures, facial expressions, and speech, their design often precludes tactile interactions due to their rigid and heavy exterior materials. In contrast, social touch, which is defined as physical contact in a social context, serves as a cornerstone of human interactions and profoundly influences both emotional and physical well-being [1]. Numerous studies have corroborated the therapeutic impact of social touch on alleviating mental and physical stress [2,3]. As social robots become increasingly more integrated into our daily lives, the importance of tactile interactions such as hugs and handshakes as a form of psychological support cannot be overstated [4,5].

To enable meaningful tactile interactions between humans and social robots, two essential capabilities must be incorporated into the design of the latter: the ability to initiate touch with humans in a socially appropriate manner as well as the capacity to respond to human touch in a human-like way upon recognition. Various tactile sensing technologies have been proposed to fulfill these requirements [6–8], including piezoelectric films and magnetic sensors embedded in silicone rubber skin [9,10] and fabric-based flexible tactile sensors [11–13]. Although these sensors perform adequately when subjected to sufficient force that deforms the skin, they struggle to detect subtler forms of touch such as gentle stroking or patting that are commonly observed in human-to-human interactions. This is primarily because these softer touches do not significantly deform the skin. Moreover, existing technologies largely concentrate on detecting interactions only after a human touch has occurred.

Research on social touch reveals that the perception of touch in interpersonal interactions is not solely determined by the tactile stimulus itself. Rather, it is significantly shaped by the cultural background, beliefs, and emotional states of the individual being touched [1]. Additionally, how a touch is initiated—specifically, the speed at which it approaches—also plays a pivotal role in its interpretation. For example, according to the concept of "Humanitude [14]," a rapidly approaching touch is generally perceived as aggressive, whereas a slower approach is deemed supportive. Indeed, several studies have indicated that the dynamics leading up to a touch—pre-touch

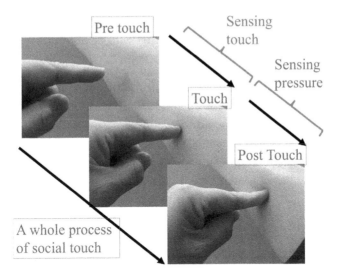

FIGURE 2.1 Three steps in social touch.

interactions—hold valuable insights for human–robot interactions (HRIs) [15,16]. This insight implies that for a robot to fully comprehend the nuances of social touch, it must be capable of recognizing human touch behaviors even before physical contact is made. This recognition cannot solely rely on visual cues, since the act of touching often occurs in areas obscured by a human or a robot body. Therefore, social robots should employ multiple, complementary methods for recognizing pre-touch interactions.

In this section, we argue that the recognition of social touch in HRIs must encompass both pre- and post-touch phases (Figure 2.1). To create a social robot that can facilitate these dual aspects of touch, we designed a sensor system that can detect both imminent and actual touch events. We first developed a proximity and touch sensor for social touch using a fabric-based, conductive sensor and described its characteristics and several applications in social touch contexts.

2.2 DESIGN CONCEPT OF A SENSOR SYSTEM FOR DETECTION OF SOCIAL TOUCH

Silvera-Tawil et al. outlined several essential criteria for touch sensors in the context of human–robot touch interactions [8]:

C1. Spatial resolutions ranging from 10 to 40 mm are generally considered adequate.

C2. Sensors should be durable, reliable, and versatile, capable of capturing a broad spectrum of data. Parallel sensing of multiple data types does not necessarily enhance performance.

C3. The force exerted in typical human touch interactions can vary from as low as 0.3 N (approximately 30 g/cm²) for a gentle stroke to over 10 N (around 1,000 g/cm²) for a push or a slap. Conventional force sensors struggle to detect subtle touches, making the incorporation of dynamic (vibration) or proximity sensors advantageous.

C4. The fabrication of the sensor skin should be straightforward, cost-effective, replicable, scalable, and result in a durable product.

C5. The sensor skin should be easily adaptable to a variety of three-dimensional robotic structures.

C6. All hardware components should be engineered to withstand human-like environmental conditions, including rapid temperature fluctuations, varying humidity levels, sudden force, stress, dust, light, and electric fields.

C7. Real-time data processing is essential. For social HRI, sampling rates between 20 and 60 Hz are deemed sufficient.

In addition to these general criteria, we propose an additional requirement:

C8. Sensors should be capable of detecting not only the touch and post-touch phases, but also the pre-touch phase. This is crucial because the impact of social touch is influenced by both the tactile stimulus and the context leading up to the touch.

To meet these criteria, we developed a sensor system for recognizing social touch that integrates two distinct functions: identifying a pre-touch to the touch phase and capturing a touch to the post-touch phase. We employed a proximity sensor for the former and a pressure sensor for the latter. Both functions were achieved in a fabric-based, conductive sensor.

2.3 FABRIC-BASED SENSOR FOR SOCIAL TOUCH

We developed a capacitive sensor based on a conductive fabric as our sensor for social touch (Figure 2.2a). The conductive fabric (Sanki Consys Co., Ltd.) is made by simultaneously knitting together

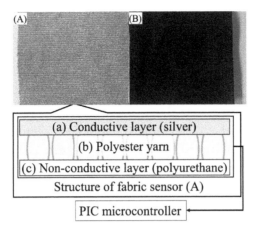

FIGURE 2.2 Fabric sensor for social touch. Fabric sensor, which is knitted withthree layers: the silver-coated yarn (a), the polyester yarn (b), and the polyethylene yarn (c).

silver-coated nylon yarn, polyethylene yarn, and polyester yarn. The fabric is knitted with three layers: the silver-coated yarn is on one side (Figure 2.2a), the polyester yarn (Figure 2.2b) is in the middle to connect both sides, and the polyethylene yarn is on the other side (Figure 2.2c). This results in a fabric that is conductive on the surface and insulative on the reverse.

In conventional fabric-type sensors [12], conductive and insulative fabrics are made separately and sewn together. In this production method, the distance between the conductive and insulating surfaces differs near the seam where tension is applied and elsewhere, resulting in variations in the sensor's sensitivity. On the other hand, in the developed fabric, the conductive and insulative surfaces are knitted at the same time, so tension is applied uniformly and the distance between the two surfaces can be kept constant to maintain reliability, meeting C2. Furthermore, the sensor is manufactured with a knitting machine, satisfying C4. Since the sensor is fabric, it is soft, pliable, and extremely robust, meeting criteria C4 and C5. Furthermore, it can be cut and sewn like a common fabric, which means that a wide variety of shapes and size can be produced, meeting criterion C1. Since all these sensor materials have been used for human cloth, we expect this sensor to satisfy criterion C6 in a social robot that coexists with humans.

The sensor is connected to a microcontroller (Microchip Technology Inc.) through a fabric-based electric wire covered with non-conductive fabric. We measured the sensor's capacitance using a measurement function implemented in the microcontroller. The gathered sensor data are sent to a PC either wired by a serial port or wirelessly in real time, satisfying C7. The sensor can be freely attached to the robot's surface using insulating cloth tape.

2.4 CHARACTERISTICS OF DEVELOPED SENSOR

We investigated whether the developed sensor had characteristics that satisfy criteria C3 and C7. First, we examined pressure changes. Figure 2.3 shows the experimental setup. We used a digital force gauge and pressed a 25-mm-diameter piece of aluminum ten times against a 5-cm-square sensor at pressures of 0.3, 1, 5, 10, 15, 20, and 25 N and measured the sensor values at those times. The results are shown in Figure 2.4. Criterion C3 proposed by Silvera-Tawil et al. requires a sensor to handle forces from 0.3 N (a gentle touch) to 10 N (a strong touch like pushing or poking). Although this sensor has sufficient sensitivity to discriminate between a gentle touch of less than 1.0 N and a strong touch of more than 20 N, the resolution is not very high. Therefore, while it can discriminate social

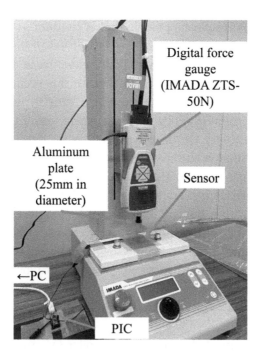

FIGURE 2.3 Experimental setting for examining basic properties of developed sensor of pressure.

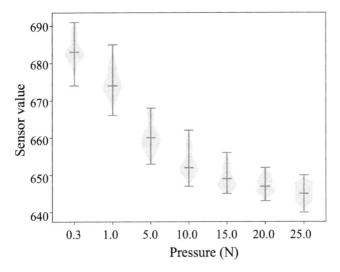

FIGURE 2.4 Sensor value changes from pressure.

touches, it is inappropriate for tasks that require fine sensing such as object manipulation.

Next, we examined the sensor's characteristics for proximity changes. Figure 2.5 shows the sensor system and an experimental setting we developed to examine its pre-touch phase characteristics. We developed a validation system in which a mannequin wore a cloth with nine 5×10 cm sensors attached to its left sleeve and measured the changes in sensor values when a person approached it. In the experiment, a male experimenter (170 cm, 80 kg) approached from a distance of 100 cm a

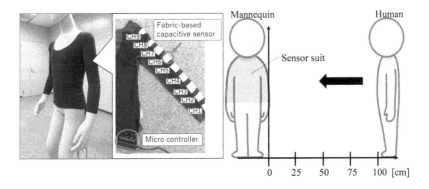

FIGURE 2.5 Developed sensor suit and experimental setting to examine its proximity properties.

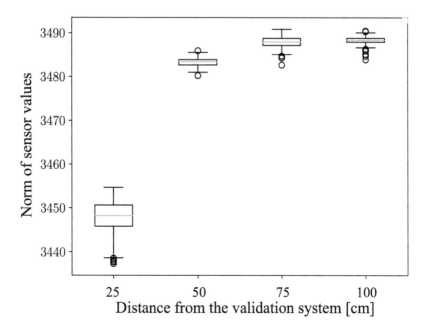

FIGURE 2.6 Sensor value changes in distance between validation system and humans.

mannequin that was wearing our developed sensor suit (Figure 2.5, right). He stopped for two minutes at 100, 75, 50, and 25 cm from the mannequin and recorded the sensor data at each distance. We computed the Euclidean norm of nine sensors for each bit of data. Figure 2.6 shows the Euclidean norms of the sensors at each distance. This result indicates that the developed system can distinguish a human from a distance of at least 75 cm, although it struggles to detect a human at more than 75 cm. This detectable distance corresponds to intimate space for embracing, touching, or whispering, as defined in proxemics [17]. We also confirmed that the developed sensor system shows significantly different values when the experimenter touched it, suggesting that its performance is sufficient to detect pre-touch and touch interactions with a human, satisfying criterion C8.

2.5 APPLICATION OF SENSOR FOR SOCIAL CONTEXT

Since the sensor we developed can be attached to clothing, it can be used to measure contact not only between people and social robots but also between people. As an example of a real-world application of our developed sensors, we have developed a system for measuring human-to-human contact during the care of seniors with dementia.

Dementia significantly impacts both seniors and their surrounding communities, including family members and caregivers. Behavioral and Psychological Symptoms of Dementia (BPSD) such as agitation, delirium, and wandering occur in 40%–50% of seniors with dementia and force caregivers to pay constant attention to the elderly; such requirements add to the burden on caregivers and increase care costs. Reducing BPSD is a major social issue in elderly care [18].

In recent years, communication with seniors with dementia is playing a more significant role in the occurrence of BPSD, and "Humanitude," a care technique that focuses on such communication, is attracting greater attention. This care technique, which is based on comprehensive communication using perception, emotion, and language, consists of 150 specific strategies based on four skills: *seeing, touching, talking,* and *standing* [19]. Due to Humanitude's effectiveness, it is being introduced in nursing care settings around the world, and the scientific community is verifying its effectiveness, analyzing it using information and communication technology, and developing support robots [20].

Although Humanitude is an effective care technique for seniors with dementia, its mastery is difficult. The communication style of seniors with dementia is very different from that of healthy adults, and a communication style must be acquired that is suitable for seniors with dementia through training. For example, in *seeing*, which is an important preliminary step to *touching* in Humanitude, caregivers are required to establish eye contact within 20 cm of a senior's face. Such an extremely close distance mocks generally accepted social norms for establishing eye contact. For *standing up*, which is done with touching, the caregiver and the cared-for person must work together appropriately while keeping their bodies in close contact to help the latter stand up on their own. However, to master these skills, the caregiver requires special training from an expert, a situation that has dampened its widespread use. If it were possible to continuously measure the proximity and contact state of caregivers during elderly care, we might be able to quantitatively evaluate Humanitude skills and construct a system that supports the effective acquisition of assistance skills even without instructors. For these purposes, developing a training support system for Humanitude has begun ([21] as an example). We developed a mask-type Humanitude training application to measure the distance

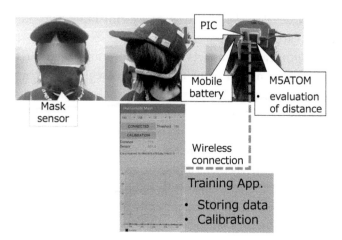

FIGURE 2.7   Overview of a mask-type Humanitude training application.

between a caregiver and a cared-for person, as well as a smock-type sensor suit to measure the state of contact during standing assistance and touching with our developed sensors.

Figure 2.7 shows an overview of our developed system, which consists of (1) a mask-cover proximity, sensor-based distance measurement device and (2) a Humanitude training application equipped with a distance judgment function. The in-facial distance measurement device and the application are connected wirelessly. The latter judges the distance based on the sensor information sent from the in-facial distance measurement device and notifies the learner by sound and vibration when the distance is within 20 cm. This allows the learner to gain an understanding of the sense of the face-to-face distance, a critical skill in Humanitude. We developed a 10-cm long, 15-cm wide mask-cover-type sensor with black insulation tape on its surface. The mask's contact surface is covered with Velcro, which was applied directly to the surface of the mask worn by the user. The sensor is connected to a microcontroller that measures the capacitance through a shielded cable for noise reduction. The measured sensor values are sent wirelessly from M5ATOM Matrix to the Humanitude Training App on a smartphone. The device can be implemented as a wearable device because it operates on a mobile battery. In this paper, all the components are put on a hat. Capacitance is inversely proportional to the distance between two persons. Therefore, the lesser the distance between people, the larger the capacitance is. When a user wearing the device approaches a person's face, the distance between the mask cover and the person's face is shortened,

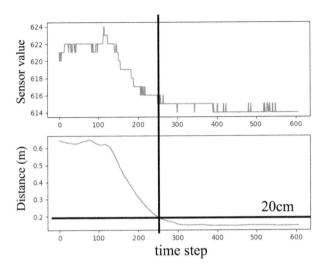

FIGURE 2.8   Example of changes in face-to-face distance and sensor values.

resulting in a larger capacitance and, consequently, a smaller sensor value from the microcontroller. This controller evaluates the distance between the mask cover and the person's face. Figure 2.8 shows an example of the change between the sensor value and the actual distance when the wearer of the face-to-face distance measurement device approaches another person's face from 60 cm away. The actual distance was measured using motion capture. The sensor value changed significantly as the mask-cover-type sensor gets closer to the person's face.

Figure 2.9 shows an overview of our developed smog-type tactile sensor suit. We selected an open-backed type since users can easily put it on and remove it off to reduce their burden. The developed suit has 33 channels of fabric-type sensors on the upper body (Figure 2.9b), including a mask cover. Each sensor is equipped with a small microcontroller board for reading capacitance, and each sensor board is wired with elastic fabric conductors to send sensor values to a transmitting microcontroller (M5ATOM Matrix) by I²C communication. The transmitter microcontroller wirelessly sends all the sensor data to PC by UDP. We prepared suits for both a caregiver and a patient to measure their contact states when the former is helping the latter to stand up. Figure 2.9c shows an example of a sensor response when a human is wearing each of the sensors and assists the person to a standing-up motion. The intensity of the red indicates the contact's strength. The sensor can detect that the caregiver is putting an arm around the assisted person's waist, that their chests are touching, and

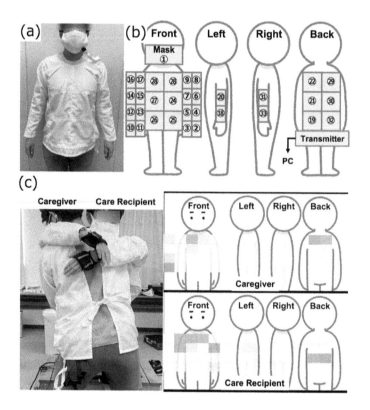

FIGURE 2.9 (a) Smock-type wearable sensor, (b) placement of sensors, and (c) example of care motion and corresponding sensor responses.

that the caregiver is putting an arm around the assisted person without enough contact with the side of the assisted person's body. By simultaneously measuring the data sent from the two sensor suits, we can confirm which body parts of the caregiver and the assisted person are in contact with each other during the measurements. We confirmed that the state of contact differed between skilled and unskilled caregivers [22]. This measurement could not be achieved by conventional sensors such as motion capture system because of large occlusion.

2.6 CONCLUSIONS AND DISCUSSION

In this section, we pointed out that the effect of social touch is determined as a whole process from pre-touch interaction to post-touch interaction and we proposed a design concept for a sensor system that can cover the whole process. First, we developed a sensor that can detect pre-touch and touch interactions using fabric-based sensor. Following the criteria to achieve social touch, we verified the properties of the developed sensor and confirmed that it has abilities to cover the whole process: discriminating between gentle and strong touches as well as capturing a human who is approaching a robot. Although research remains immature on social touch in HRI from pre- to post-touch interactions, our system helps a social robot improve the robustness of perception at close distance.

Since the developed sensor is wearable, we can apply it not only for robots but also for humans. Therefore, we can also use it in investigations of human–human touch interaction. We showed two such applications for elderly care. This investigation sheds light on the understanding of the whole procedure of social touch and sensor specifications such as their placement and sizes.

Finally, we point out the individual variability in terms of the electrostatic capacity of the human body and the change of capacity from long-term use. We expect to reduce these problems by calibration for each individual and measuring skin conductance. Future work will implement our developed sensor suit in a humanoid robot to investigate the effect of its motion on the sensor data. We have already started this work [23]. The integration of more sensors is also critical.

ACKNOWLEDGMENT

This research work was supported in part by JST CREST Grant Numbers JPMJCR18A1 and JPMJCR17A5, Japan.

REFERENCES

[1] A. Gallace and C. Spence: "The science of interpersonal touch: An overview," *Neuroscience and Biobehavioral Reviews*, 34(2), pp. 246–259, (2010).

[2] T. Field: "Touch for socioemotional and physical well-being: A review," *Developmental Review*, 30, pp. 367–383, (2010).

[3] S. Cohen: "Social relationships and health," *American Psychologist*, 59, pp. 676–684, (2004).

[4] C. J. A. M. Willemse, A. Toet, and J. B. F. van Erp: "Affective and behavioral responses to robot-initiated social touch: Toward understanding the opportunities and limitations of physical contact in human-robot interaction," *Frontiers in ICT*, 4, p. 12, (2017).

[5] K. Nakagawa, M. Shiomi, K. Shinozawa, R. Matsumura, H. Ishiguro, and N. Hagita: "Effect of robot's active touch on people's motivation," In *Proceedings of the 6th ACM/IEEE International Conference on Human-robot Interaction*, pp. 465–472, (2011).

[6] R. S. Dahiya, P. Mittendorfer, M. Valle, G. Cheng, and V. Lumelsky: "Directions towards effective utilization of tactile skin—A review," *IEEE Sensors*, 13(11), pp. 4121–4138, (2013).

[7] B. D. Argall and A. G. Billard: "A survey of tactile human-robot interactions," *Robotics and Autonomous Systems*, 58(10), pp. 1159–1176, (2010).

[8] D. Silvera-Tawil, D. Rye, and M. Velonaki: "Artificial skin and tactile sensing for socially interactive robots: A review," *Robotics and Autonomous Systems*, 63(3), pp. 230–243, (2015).

[9] T. Noda, T. Miyashita, H. Ishiguro, K. Kogure, and N. Hagita: "Detecting feature of haptic interaction based on distributed tactile sensor network on whole body," *Journal of Robotics and Mechatoronics*, 19(1), pp. 42–52, (2007).

[10] T. Kawasetsu, T. Horii, H. Ishihara, and M. Asada: "Flexible Triaxis tactile sensor using a spiral inductor and magnetorheological elastomer," *IEEE Sensors*, 18(14), pp. 5834–5841, (2018).

[11] H. Alirezaei, A. Nagakubo, and Y. Kuniyoshi: "A highly stretchable tactile distribution sensor for smooth surfaced humanoids," In *Proceedings of the IEEE-RAS International Conference on Humanoid Robots*, pp. 167–173, (2007).

[12] M. Inaba, Y. Hoshino, K. Nagasaka, T. Ninomiya, S. Kagami, and H. Inoue: "A full-body tactile sensor suit using electrically conductive fabric and strings," In *Proceedings of the IEEE/RSJ International Conference on Intelligent Robots and Systems*, 2, pp. 450–457, (1996).

[13] G. H. Büscher, R. Kōiva, C. Schürmann, R. Haschke, and H. J. Ritter: "Flexible and stretchable fabric-based tactile sensor," *Robotics and Autonomous Systems*, 63, pp. 244–252, (2015).

[14] Y. Gineste and J. Pellissier: *Humanitude, cuidar e compreender a velhice.* Lisboa, Portugal: Instituto Piaget, (2008).

[15] M. Shiomi, K. Shatani, T. Minato, and H. Ishiguro: "How should a robot react before people's touch?: Modeling a pre-touch reaction distance for a robot's face," *IEEE Robotics and Automation Letters*, 3(4), pp. 3773–3780, (2018).

[16] S. Keshmiri, M. Shiomi, K. Shatani, T. Minato, and H. Ishiguro: "Facial pre-touch space differentiates the level of openness among individuals," *Scientific Reports*, 9(1), p. 11924, (2019).

[17] E. T. Hall: *The Hidden Dimension.* Knopf Doubleday Publishing Group: Anchor Books, (1966).

[18] G. Livingston et al.: "Dementia prevention, intervention, and care," *The Lancet*, 390(10113), pp. 2673–2734, (2017).

[19] Y. Gineste and J. Pellissier: *Humanitude : Comprendre la vieillesse, prendre soin des Hommes vieux.* Aramand Colin, (2007), doi:10.4000/adlfi.7588.

[20] H. Sumioka, M. Shiomi, M. Honda, and A. Nakazawa: "Technical challenges for smooth interaction with seniors with dementia: Lessons from humanitude™," *Frontiers in Robotics and AI*, 8, p. 650906, (2021), doi:10.3389/frobt.2021.650906.

[21] A. Nakazawa, M. Iwamoto, R. Kurazume, M. Nunoi, M. Kobayashi, and M. Honda: "Augmented reality-based affective training for improving care communication skill and empathy," *PLoS One*, 18(7), e0288175, (2023), doi:10.1371/journal.pone.0288175.

[22] Q. An, A. Tanaka, K. Nakashima, H. Sumioka, M. Shiomi, and R. Kurazume: "Understanding humanitude care for sit-to-stand motion by wearable sensors," In *2022 IEEE International Conference on Systems, Man, and Cybernetics (SMC)*, pp. 1874–1879, (2022), doi:10.1109/SMC 53654.2022.9945156.

[23] Y. Onishi, H. Sumioka, and M. Shiomi: "Moffuly-II: A robot that hugs and rubs heads," *International Journal of Social Robotics*, (2023), doi: 10.1007/s12369-023-01070-5

Application of Fabric Sensors for Soft Robot Hand for Positioning an Object without Touching it

Takashi Takuma, Koki Haruno,

Hidenobu Sumioka, and Masahiro Shiomi

Osaka Institute of Technology, Osaka, Japan

Advanced Telecommunications Research Institute International, Kyoto, Japan

3.1 INTRODUCTION

Soft robots made of materials such as silicone and rubber have attracted much attention in recent years. By exploiting their flexibility, soft robot technology provides a wide range of applications: surgery [1], infrastructure research [2], and food preparation [3]. Soft robotic hands can touch people and food with a low possibility of injury or damage [4]. Yoshiyuki et al. developed a wrapping hand that can grasp a small amount of food, such as chopped leeks and edible seaweed [5]. Navas et al. developed soft grippers for gripping fragile fruits or vegetables [6]. Mimori et al. developed a robot hand that uses a binding mechanism

DOI: 10.1201/9781003384274-5

for grasping [7]. In factories, soft robotic hands have packed cake and grabbed fried foods. As described above, a wide variety of soft robotic hands have grasped various objects. However, these soft robotic hands are assumed to be set at a position from which they can grasp, and no method has been established yet that moves a robot's hand toward the object that will be grasped (positioning). Positioning a robot hand is not an inherent issue for soft robots; this issue applies to rigid robot hands. Some rigid robot hands have sensors, including time-of-flight sensors or cameras, attached to their fingertips to measure an object without touching it. This approach provides a more compact setup than one in which the camera is situated outside the robot. Hasegawa et al. attached a net-like proximity sensor to a robot hand's fingertip to control its posture [8]. On the other hand, for soft hand robots, conventional rigid sensors inhibit the deformation of their hands. Therefore, a sensor that positions a soft robot hand is required to accommodate the robot body's deformation.

In this study, we used a flexible conductive cloth [9] covered with silicone as a sensor. The cloth does not prevent the deformation of the soft robot hand, and its capacitance is influenced by the distance to a conductive object. Therefore, we believe that we can measure the distance between the robot hand and a grasping candidate without touching it. To observe the influence of the distance, we developed a plate-shaped sensor in which the cloth is covered by silicone to form a relationship between the capacitance of the cloth and the distance to the conductive or non-conductive object. Based on the result, we set two plate-shaped sensors, assuming that two pieces of cloth are embedded in the fingertips of the soft robot hand, and we estimated the distance between the object's center position and the middle point of sensors by detecting the moment when the highest capacitance is measured. Finally, we developed a soft robot hand, embedded cloth in its fingertip, and estimated the grasping candidate's center position. The robot also estimated the position of materials such as wood, a steel can, and curing tape and compared their estimation accuracies.

3.2 BASIC CHARACTERISTICS OF CAPACITIVE CLOTH

We adopted a capacitive cloth (Sanki Consys) (Figure 3.1a) to achieve non-contact information as the distance between the cloth and an object. The cloth and its structure are shown in Figure 3.1b. The former has a conductive layer and a non-conductive layer interwoven into an

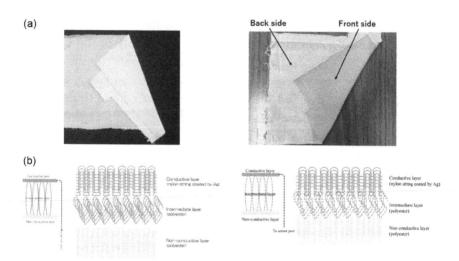

FIGURE 3.1   (a) Conductive cloth and (b) its configuration made of two layers: a conductive layer woven by nylon string coated by silver and a non-conductive layer woven by polyester. Layers are connected by intermediate layer woven by polyester. Capacitance is measured by connecting lead wire to conductive layer.

intermediate layer. The conductive surface consists of nylon silver-plated yarn. The non-conductive surface consists of polyester. The conductive and non-conductive surfaces are connected by polyester. Because the conductive layer works as a polar plate, the capacitance is determined by the distance between the cloth and an object to be grasped.

To observe the change in capacitance caused by conductive objects, a plate-shaped sensor was made as a prototype. A lead wire is attached to the conductive cloth using a reinforcing cloth (Figure 3.2a) and shown in Figure 3.2b. Assuming that the cloth is embedded in a soft robot hand made of silicone, it is covered with silicone (Ecoflex 00-30). The fabricated sensor (Figure 3.2c) is formed by pouring silicone solvent into a mold made by a 3D printer. The conductive cloth is $10 \times 50 \times 1$ mm; the sensor is $20 \times 70 \times 4$ mm. The thickness between the surfaces of the conductive cloth and the silicone covering (the sensor's surface) is 1 mm.

The capacitance was measured using a microcontroller (Arduino). Figure 3.3 shows a circuit that measured the capacitance change. The time was measured when the voltage of the receiving pin exceeded a particular threshold (the charging period). The larger the sensor's capacitance is, the longer it takes for the receiving pin to exceed the threshold. Therefore, the charging period determines the sensor's capacitance.

FIGURE 3.2 Developed sensor made of conductive cloth covered by silicone: (a) connection with lead wire using reinforcement cloth and (b) cloth connected by lead wire. (c) Plate-shaped sensor where cloth is covered by silicone.

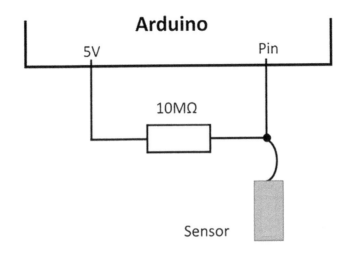

FIGURE 3.3 Circuit that measures sensor's capacitance: Period until pin detects high voltage is counted.

We recorded the charging period when an object was near the sensor. Two types of objects were selected: wood as an insulator and aluminum lumps as a conductor. The sizes of both were identical: 30-mm wide, 15-mm long, and 100-mm high. As shown in Figure 3.4, the object was placed above the sensor and approached from 50 mm in 5-mm increments to the shortest distance without contact of 1 mm and with contact of 0 mm. For noise reduction, the average values were recorded every ten seconds.

Figure 3.5a and b shows the wood and aluminum results. The vertical axis shows the charging period and its positive correlation with the capacitance, and the horizontal one shows the distance between the sensor and the object. When wood was the approaching object (Figure 3.5a), the

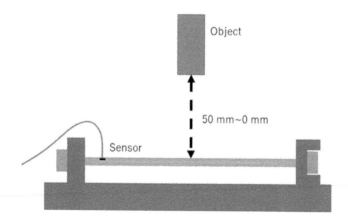

FIGURE 3.4 Experiment setup that measures charging period when object is approaching sensor.

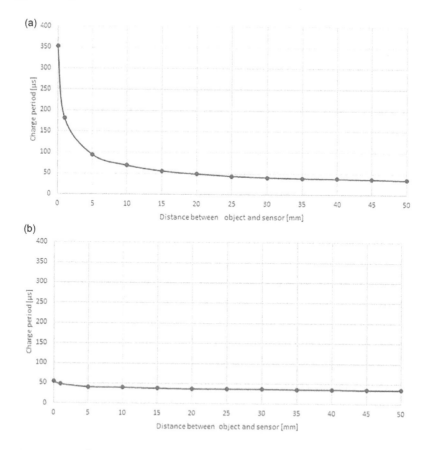

FIGURE 3.5 Charging periods of sensor: (a) wood is a non-conductor and (b) aluminum is a conductor approaching sensor.

charging period only changed slightly when the object it approached to the sensor was within 1 mm and made contact with the sensor. In the case of aluminum (Figure 3.5b), the charging period greatly changed when the distance between the sensor and aluminum was less than 20 mm, although a larger change was observed when the object made contact with the sensor. Utilizing this characteristic, the next section explains the estimation of the grasping candidate without touching it using two plate-shaped sensors for positioning the soft robot hand.

3.3 ESTIMATION OF POSITION OF TARGET OBJECT

The robot hand in Figure 3.6a can successfully grasp the object when it is set in the middle of the fingers, i.e., the center of its hand should coincide with the object's center, assuming that the fingers are symmetrically open. In this section, supposing that the robot's fingertips into which the cloth is embedded move over the object without touching it in the next subsection, the position of the object's middle that corresponds to the middle of the sensor as shown in Figure 3.6a is estimated using two plate-shaped sensors.

Because the distance changes between the fingertip and the object due to the finger's fluctuation from the softness of the soft finger when the hand moves over the object, this section adopts fixed two plate-shaped sensors with a certain distance and the object is moved under the sensors (Figure 3.6b) to check the accuracy of the estimated distance. The distance between the object's middle and the middle position of the two sensors is estimated by measuring the capacitance of each one. As

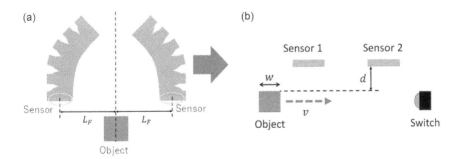

FIGURE 3.6 (a) Positioning of robot hand by moving over object and measuring the capacitance of sensors embedded in the fingertips. (b) Testbed to check the accuracy of position estimation. Plate-shaped sensors were fixed, and the object moved.

explained in the previous section, the capacitance changes when the object is conductive. Therefore, the aluminum lump used in the previous section was adopted as object.

For the estimation of the distance, the object was moved under the sensors (Figure 3.7), and we recorded the times when the capacitance of each sensor reached its maximum. Assuming that the robot hand's moving velocity is controllable, the object's velocity in this section v was given. A push switch detected a terminal of the moving object, which recorded the following times: when the object moved from its arbitrary position (Figure 3.7a); when the capacitance of sensor 1 reached its maximum T_{M1}

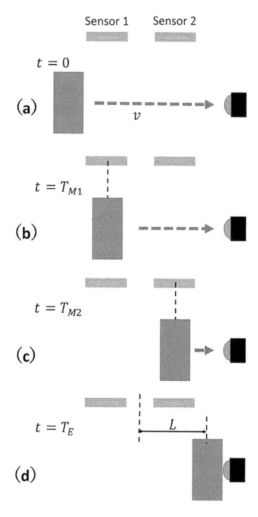

FIGURE 3.7 Procedure that estimates distance between middle position of two sensors and object's middle L.

(Figure 3.7b); when the capacitance of sensor 2 reached its maximum T_{M2} (Figure 3.7c); and when its right side touched the surface of switch T_E (Figure 3.7d). From the velocity and the recorded times, distance L from the object's middle to the middle position of the sensors is calculated as

$$L = \left(T_E - \frac{T_{M1} + T_{M2}}{2} \right) v,$$

where v is the velocity of the object. By moving the object a distance L, the middle position of two sensors will coincide with the object's middle.

Figure 3.8 shows the experimental setup. Since the slider moved up and down, the sensors were placed vertically and the object attached to the slider moved from top to bottom. We assumed no gravitation effect during our estimation. The object stopped when it contacted the switch. The slider moved at $v = 5$ mm/s. To investigate the relationship among distance d, the length between the sensor and the object (Figure 3.6b), and the estimation accuracy, d was varied from 1, 5, 10, 15, and 20 mm, and the position estimation error was recorded at each distance. To smooth out the noise, the charging period was the average of the last ten records.

Figure 3.9 shows an example of the change in the charging period when a conductor passes between two sensors. The vertical axis shows the charging period, while the horizontal axis shows the time. The charging periods reach their highest value when sensors 1 and 2 are about 11,000 and 25,000 ms. The error and the relative error between calculated position L and measured value 112.5 mm for each distance d are shown in Table 3.1. For each distance, L was estimated ten times.

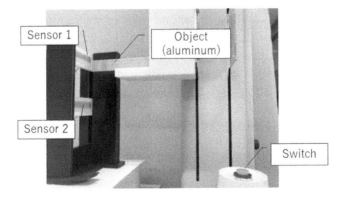

FIGURE 3.8 Experimental setup: Sensors are placed vertically, and object moved down assuming no gravitational effect on the estimation. It stops when it touches the switch.

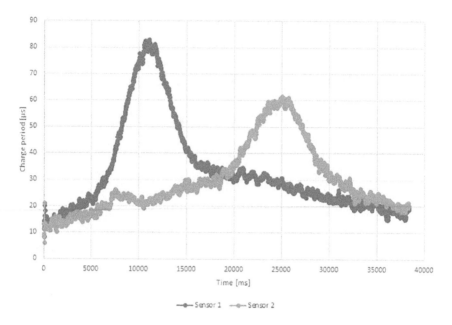

FIGURE 3.9 Example of charge periods of both sensors while an object moves over sensors. The time when the charge period is the highest indicates the minimum distance between the object and the sensor.

TABLE 3.1 Experimental Results Using Two Plate-Shaped Sensors

Distance d [mm]	1.0	5.0	10.0	15.0	20.0
Average error [mm]	4.5	4.3	3.3	4.5	10.1
Relative error [%]	4.0	3.9	2.9	4.0	9.0

The results show that the average and relative errors were 10 mm and 9.0% when d was 20 mm. When d is 15 mm or less, there is no significant difference in the average error. Considering that the width of object is 15 mm and the fingers can open much wider than that, the error is negligible.

3.4 POSITION ESTIMATION USING TWO SOFT ROBOT FINGERS

The previous section showed that the distance between object's middle and the middle position of the sensors can be estimated by moving the object over the sensors and recording the highest value of each sensor's charging period. This section explains the development of a soft robot hand that is composed of two soft fingers, and we experimentally estimated the object's position using two conductive cloths embedded in the soft robot's fingertip. Soft robot fingers were formed by pouring silicone solution into a mold made by a 3D printer in the same manner as with the plate-shaped sensors. The developed soft robot fingers are shown in Figure 3.10a. A finger has a cavity that stores air. When it is filled with air of a certain pressure, it deforms and achieves a C-shaped curve (Figure 3.10b). The robot has two fingers, and conductive cloth is embedded in the fingertips (Figure 3.10c). As explained above, the fingers symmetrically open, and the hand's center position matches the sensor's middle point. Because the distance between the sensors changes based on the posture of the fingers and is difficult to measure, we didn't use the distance between the sensors for the estimation as in the previous section.

As shown in Figure 3.11, the robot hand moved over the object from left to right at a constant speed and stopped after certain period T_{hE}. Similar to the method mentioned in the previous section, distance L_h between the terminal position of the hand to the center of the object is estimated by the following equation:

$$L_h = \left(T_{hE} - \frac{T_{hM1} + T_{hM2}}{2} \right) v$$

where T_{hM1} is the time when the charging period of sensor 1 (the right sensor in the figure) reaches its maximum value. T_{hM2} is the time when the charging period of sensor 2 (the left sensor) reaches its maximum value.

The experimental setup is shown in Figure 3.12. The hand was carried by a slider driven by a stepping motor. Figure 3.12a shows the hand's initial position, whereas Figure 3.12b shows its terminal position. As shown in Figure 3.11, the actual distance from the center of the object

FIGURE 3.10 Developed robot hand that contains two soft silicone fingers: (a) developed soft robot finger and (b) curved form when filled with air. (c) Soft robot hand equipped with two soft robot fingers and conductive cloth is embedded in the fingertip.

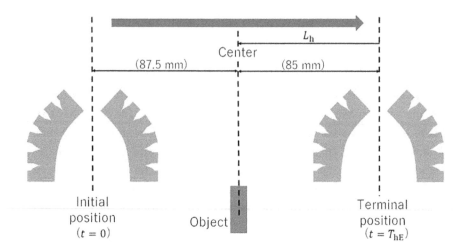

FIGURE 3.11 Motion of robot hand to estimate center of grasped object. Note that time to stop hand T_{hE} is known.

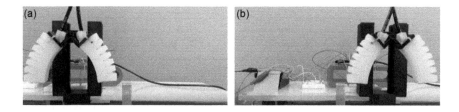

FIGURE 3.12 (a) Experimental setup and (b) hand movement.

to the terminal position is 85 mm, i.e., if L_h is estimated as 85 mm, the estimation is correct. The capacitance does not increase if the distance between the fingertip and the object is long (Figure 3.5b). On the other hand, if the distance is too close, the fingertip touches the object by the fluctuation of the soft finger as it moves, even though this is not the time when the sensor is closest to the object and the capacitance is too high. Therefore, we set the distance to 3 mm so that the fingertip did not touch the object since the sensor's fluctuation and capacitance increase with more distance from the object. The charging period was the average of the previous five points.

Figure 3.13 shows an example of the change in the charge periods of the right and left sensors (sensors 1 and 2) as the robot hand moved. Similar to the previous experiment, the period of sensor 1 first reached its maximum value and then the period of sensor 2 reached its maximum value.

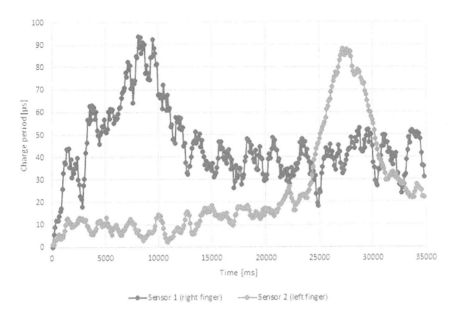

FIGURE 3.13 Example of charge periods of right and left sensors. Trajectory fluctuated due to finger's softness.

TABLE 3.2 Estimated Distance L_h and Error

Trials	1	2	3	4	5
Error [mm]	0.1	0.4	2.6	0.4	1.8
Relative error [%]	0.1	0.5	3.0	0.5	2.1

Unfortunately, due to the fluctuation of the soft robot finger, cyclic fluctuation was observed. Table 3.2 shows the estimated positions and the relative errors of five trials. The maximum error was 2.6 mm compared to a measured value of 85 mm, indicating that estimation is possible with small error. Position estimation by a soft robot hand without touching is possible with small position error estimation, although the capacitance fluctuation may be large due to the soft material.

We verified the effectiveness of the proposed method for grasping objects other than aluminum. For grasping objects, we selected a piece of wood (Figure 3.14, 9-mm wide, 70-mm deep, 125-mm high), a steel can (a 52-mm diameter, 104-mm high), and curing tape, which is a polyester tape coated with polyethylene (49-mm wide, a 98-mm diameter). As in the previous experiment, the position was estimated five times for each object, and the error and relative error were calculated. The location of the

FIGURE 3.14 Objects made of different materials: (a) wood, (b) steel can, and (c) curing tape coated with polyethylene.

TABLE 3.3 Estimated Distance L_h and Error for Wood

Trials	1	2	3	4	5
Error [mm]	1.1	3.1	21.3	19.8	4.0
Relative error [%]	1.3	3.7	25.1	23.3	4.7

TABLE 3.4 Estimated Distance L_h and Error for Steel Can

Trials	1	2	3	4	5
Error [mm]	2.0	0.2	0.9	3.5	2.0
Relative error [%]	2.4	0.2	1.1	4.1	2.4

grasping candidate was the same as in the previous experiment: 85 mm between the middle point of sensors at the hand's terminal position and the center of the object. The results for wood as an insulating material were less accurate than the previous results (Table 3.3). Accurate position estimation was difficult because the capacitance change was small when the sensor approached the wood, and the maximum value was sometimes wrongly detected. On the other hand, the capacitance with steel changed largely when the sensor reached the can at its closest distance, and an accurate position estimation was possible (Table 3.4). The capacitance also changed for the curing tape. Although a relative error of about 10% was observed in some trials, position estimation is possible, compared with

TABLE 3.5 Estimated Distance L_h and Error for Curing Tape

Trials	1	2	3	4	5
Error [mm]	1.1	1.7	8.5	7.8	8.7
Relative error [%]	1.3	1.8	10.0	9.1	10.2

the result with wood because the error was less than 10 mm (Table 3.5). Perhaps the tape's material, polyethylene, was electrically charged and generated capacitance between the sensor and the tape.

3.5 CONCLUSION

We developed a flexible sensor whose capacitance changes in response to the approach of a conductive material for which a soft robot hand achieves positioning for an object to be grasped. We investigated the sensor's characteristics and confirmed that its position can be estimated by the plate-shaped sensor that embedded the conductive cloth. The flexible sensor's characteristics showed no significant change in capacitance when the insulating material (wood) was close to the sensor. Capacitance with a conductive material (aluminum lumps) changed largely when the distance was less than 20 mm. In the position estimation using two plate-shaped sensors, the error was reduced when the distance between the sensor and object was less than 20 mm. In experiments where the cloth was embedded in a soft robot hand, we confirmed that the error was small and estimation was possible without touching the object, although we did observe capacitance fluctuation due to the soft robot fingers. In position estimation using different materials, we experimentally found that estimation was possible with a material other than metal: curing tape. Our experiments confirmed that a sensor that does not interfere with the deformation of a soft robot hand can be fabricated and that position estimation can be performed without the support of external sensors such as a camera. However, position estimation is impossible if the object to be estimated is an insulator, and the distance between the sensor and an object must be less than 20 mm. Such problems can probably be solved by selecting a material that is more sensitive to the conductor.

ACKNOWLEDGMENT

This study was partly supported by JST CREST Grant Number JPMJCR18A1 and JSPS KAKENHI Grant Number 20K11915 and 19H04193 Japan.

REFERENCES

[1] M. Runciman, A. Darzi, and G. P. Mylonas, "Soft Robotics in Minimally Invasive Surgery," *Soft Robotics*, vol. 6, no. 4, pp. 423–443, 2019.

[2] E. Milana, "Soft Robotics for Infrastructure Protection," *Frontiers in Robotics and AI*, vol.9, https://www.frontiersin.org/articles/10.3389/frobt.2022.1026891, 2022.

[3] Y. Yamanaka, S. Katagiri, H. Nabae, K. Suzumori, and G. Endo, "Development of a Food Handling Soft Robot Hand Considering a High-Speed Pick-and-Place Task," *2020 IEEE/SICE International Symposium on System Integration (SII)*, pp. 87–92, 2020.

[4] R. Liu, J. Hao, X. Li, H. Su, and C. Shi, "A Soft Pneumatic Gripper Integrated with a Flexible Capacitive Pressure Sensor," *IEEE International Conference on Robotics and Biomimetics*, pp. 833–838, 2021.

[5] Y. Kuriyama, Y. Okino, Z. Wang, and S. Hirai, "A Wrapping Gripper for Packaging Chopped and Granular Food Materials," *2019 2nd IEEE International Conference on Soft Robotics (RoboSoft) COEX*, pp. 114–119, 2019.

[6] E. Navas, R. Fernandez, D. Sepulveda, M. Armada, and P. Gonzalez-de-Santos, "Soft Grippers for Automatic Crop Harvesting: A Review," *Sensors*, vol. 21, no. 8, p. 2689, 2021.

[7] Y. Mimori, Z. Wang, and S. Hirai, "A Novel Binding Hand with Closed Loop Thread Capable of Grasping Small-Diameter Objects," *2019 IEEE International Conference on Soft Robotics* (RoboSoft 2019), pp. 310–315, 2019.

[8] H. Hasegawa, Y. Suzuki, A. Ming, K. Koyama, M. Ishikawa, and M. Shimojo, "Net-Structure Proximity Sensor: High-Speed and Free-Form Sensor With Analog Computing Circuit," *IEEE/ASME Transactions on Mechatronics*, vol. 20, no. 6, pp. 3232–3241, 2015.

[9] T. Takuma, K. Haruno, K. Yamada, H. Sumioka, T. Minato, and M. Shiomi, "Stretchable Multi-Modal Sensor Using Capacitive Cloth for Soft Mobile Robot Passing through Gap," *IEEE International Conference on Robotics and Biomimetics*, pp. 1960–1967, 2021.

CHAPTER 4

Wearable Tactile Sensor Suit for Natural Body Dynamics Extraction

Case Study on Posture Prediction Based on Physical Reservoir Computing

Hidenobu Sumioka, Kohei Nakajima,

Kurima Sakai, Takashi Minato,

and Masahiro Shiomi

Advanced Telecommunications Research Institute International, Kyoto, Japan

The University of Tokyo, Tokyo, Japan

RIKEN, Saitama, Japan

4.1 INTRODUCTION

As robotics technology continues to progress, interest is growing in the tactile aspects of human–robot interactions such as object manipulation and touching. Many current sensors are rigid and intended to be attached to the hard surface of a robot's body. However, such sensors cannot be used on soft-skin robots such as android robots that closely resemble humans

DOI: 10.1201/9781003384274-6

and stuffed robots. In prior research, we introduced a tactile sensory suit tailored for these anthropomorphic androids [1, 2].

In many previous robots, each sensor was arranged to avoid shared or mutual interference, and sensor responses were often separated from the robot's motions. On the other hand, in our proposed tactile sensor suit, the suit itself changes based on the robot's motion, so that the robot's motion information is reflected in the sensor values. This means that dynamics are present among the tactile sensor, the clothing, and the robot's motion, and the clothing becomes a large-scale sensor network that reflects the robot's motion.

Research on physical reservoir computing has been burgeoning in the current scholarly landscape. This approach regards the dynamic behavior of soft materials as computational devices like recurrent neural networks, highlighting the potential for diverse physical phenomena to serve as computational resources [3, 4]. A subset of these studies has focused on leveraging a robot's body as a computational resource, investigating the computational prowess of musculoskeletal systems [5], or generating quadruped robot locomotion using a soft spine as a physical reservoir [6]. Extending the boundaries, some research has even engaged soft-bodied creatures such as an octopus for information processing [7–10].

These studies hint at the potential for active applications of tactile sensor suits to monitor inherent body dynamics. Monitored dynamics could be directly harnessed for information processing to estimate the state of android robots. If the tactile sensors embedded in the suit can predict robotic movements, then android robots might maintain their robust performance by the self-assessment of their posture using only tactile feedback from removable clothing, even in the event of malfunctioning internal sensors, which are harder to replace.

In this section, we extend the concept of physical reservoir computing to fabric-based tactile sensors and propose an innovative fabric-based information processing apparatus that employs clothing and its wearer as computational resources. We treat a sensor suit furnished with an array of touch sensors as a recurrent neural network and use time series data collected from the sensors to estimate the robot's movements or the person wearing the suit. By directly engaging a sensor suit laden with many tactile sensors as an information processing tool, processing can be executed proximally to the user or the robot. This feature presents an intriguing prospect for edge computing [11] with minimized latency. In this section, we validate this concept by exploring the degree to which robot movements can be estimated utilizing a simple fabric tactile sensor suit.

4.2 TACTILE SENSOR SUIT FOR EXTRACTING A COMPUTATIONAL RESOURCE

Figure 4.1 shows both the conventional and our proposed approaches for the sequence learning of robot behavior. In the former, externally added machine learning systems with high computational power such as recurrent neural networks mainly take over the nonlinear and temporal problem of learning a certain target signal using sensor values. On the other hand, in our approach, we fully exploit many tactile sensors together with the wearer's dynamics, which resembles a recurrent network, to solve tasks requiring nonlinearity and memory by just adding linear and static readouts. This is useful for tasks that require robots to respond in real time, such as social interactions, since it drastically reduces learning time and computational cost.

4.3 EXPERIMENTS

We conducted a preliminary experiment that explored the feasibility of estimating a robot's state through fabric-based information processing. This experiment estimated a wearer's periodic movements using a

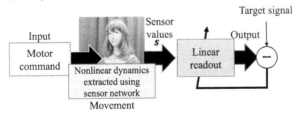

FIGURE 4.1 Typical sequence learning approach (a) and the proposed approach (b). The conventional approach represented in (a) not only exploits the sensor network as a computational resource, but also uses nonlinearity and memory incorporated from external machine learning networks at the readout. Our approach utilized the worn touch sensor network by a robot as a computational resource to solve nonlinear and temporal information processing for time series data [12].

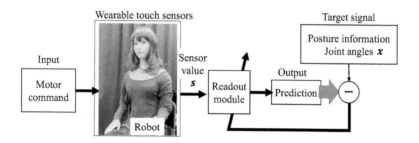

FIGURE 4.2   Schematic overview of the conducted experiments.

sensor suit (Figure 4.2). An android robot, equipped with this sensor suit, carried out a variety of motions. Subsequently, these movements were predicted, drawing on information gleaned from clothing sensor s and posture data x collected from the joint angle sensor embedded within the robot.

4.3.1 Experimental Setup

4.3.1.1 Wearable touch sensor suit

1. In our experimental setup, the tactile sensor suit is equipped with nine sensors on its left arm (Figure 4.3). Each sensor is connected to a PIC microcontroller (PIC16F1847, Microchip Technology Inc.), facilitated through a fabric-based electrical wire encapsulated within non-conductive fabric. The capacitance of each sensor was measured utilizing a function built into the microcontroller.

    The data collected from the sensors are relayed to the host PC by a USB bus and a serial port at 100-ms intervals, with the help of a capacitance calculation function appended to the microcontroller. The interface between the sensor and the human or robot body is coated with insulating cloth tape, establishing a non-zero contact distance between the sensor and the wearer.

    The wearer's posture is mirrored in the variations in contact states between the sensors and the wearer's body. For instance, when the wearer flexes her elbow, the sensors in proximity to it establish close contact with the wearer's body, thereby increasing the capacitance of the sensors. Conversely, when the wearer extends her elbow, the sensors near it either lose contact with or only partially touch her body, lowering the sensor values than when the elbow is flexed.

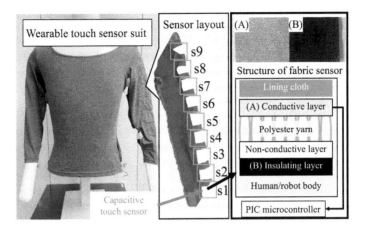

FIGURE 4.3   Wearable touch sensor suit that consists of nine capacitive touch sensors [12].

4.3.1.2 Information processing in the proposed architecture:

2. Figure 4.2 illustrates the information processing flow in our proposed system. The processing is divided into two phases: training
and testing. In the training phase, an android robot wearing the sensor suit performs a specific action for approximately 80 seconds. We
collected time series data of the sensor suit's sensor values and the
wearer's action data. Out of 600 samples extracted from these collected data, we utilized the first two-thirds (400 samples) for training
and the remainder for testing. We designed readout modules that
predicted the wearer's movement data, taking the sensor suit's current step sensor values s_t as input, and next step's movement data x_{t+1}
as target signals. The task is to learn function f such that $x_{t+1} = f(s_t)$.
We used the remaining data (200 samples) to test and evaluate
the performance of the readout modules by calculating the mean
squared error (MSE) between output $\hat{y}$ of the readout module and
actual motion data y. For all the training processes, the input and
output data were standardized between 0 and 1.

Our study explores the capabilities of a fabric-based computer:
Does the sensor time series of the sensor suit hold the dynamics to
estimate the wearer's motion trajectories? To address this question,
we utilized linear regression (LR) as the readout module to mimic the
motion trajectories in subsequent steps. In other words, we learned
the weights of the linear model: $x_{t+1} = w_0 + w_1 s_{1,t} + w_2 s_{2,t} + \cdots + w_9 s_{9,t}$,

where $s_{i,t}$ represents the sensor value of the i-th sensor (Figure 4.3) at time step t and w_i denotes its weight. For comparison, we also evaluated the performance of three recurrent neural networks for learning time series data: echo state networks (ESN) [13], long short-term memory networks (LSTM) [14], and gated recurrent unit networks (GRU) [15]. An ESN is a recurrent neural network whose learning fine-tunes the linear and static readout weights, producing quicker and more stable implementation and training compared to standard backpropagation-through-time algorithms used in LSTM and GRU [16]. In these networks, we varied the number of nodes in the network (equivalent to the number of sensors): 9, 20, 100, and 500. In ESN, we set the leaking rate and the spectral radius to 0.7 and 0.9, respectively, and employed pseudo-inverse matrix calculation for training. For LSTM and GRU, the loss function was predicated on the MSE between the network's outputs and target signals. We utilized the adaptive moment estimation method as an optimization scheme with previously recommended parameters [17] and updated the weights with training data over many iterations (epochs) to train LSTM and GRU. Training was repeated until the model had completed 1,000 epochs or the difference between two consecutive values of the loss function fell below threshold θ. We set $\theta = 10^{-5}$. For ESN, LSTM, and GRU, we abstained from using any regularization techniques. All the information processing systems were constructed in Python 3.7. LR was implemented with the sklearn module, ESN with the easyesn module, and LSTM and GRU with the pytorch module.

4.3.2 Experiment: Emulation of Robot Motion
4.3.2.1 Experimental setting

3. In our experiment, we equipped ERICA (a female android robot) with a sensor suit to ascertain how efficiently our proposed system estimated the state of the joint angle sensors during the robot's motion. ERICA possesses a total of 44 degrees of freedom (DOF) in her eyes, face, and upper body, all of which are driven by pneumatic actuators except for her eye movements. We collected the position information for each actuator within a range of 0–255. The experiment focused on estimating the sensor values of nine DOFs (Figure 4.4) with the sensor data acquired from the sensor suit: the linear displacements of

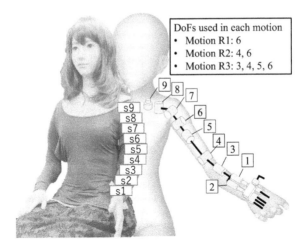

FIGURE 4.4   Degrees of freedom used in ERICA. Third to sixth joints generate motions R1, R2, and R3 [12].

joints 1, 2, and 9 (linear actuators) and the rotational displacements of joints 3–8 (rotational actuators) were evaluated as outputs of the readout modules.

In the experiment, ERICA repeatedly executed three distinct motions: (R1) swinging back and forth with a bent arm, (R2) extending her hand outward, and (R3) twisting it. During each motion, the control values were adjusted solely for the joints represented in Figure 4.4, while the others were held at a constant value. For each motion, the robot performed for about 80 seconds, and the data were gathered over five iterations.

4.3.2.2 Result

4. Figure 4.5 presents the average MSE in tests conducted on all the motion data, segmented by the number of network nodes. Note that since the number of network nodes is not a parameter on the LR, it maintains a consistent representation across all graphs. Given our training conditions, LR, which relies on a weighted linear sum of sensor values from the sensor suit, demonstrates performance on par with recurrent network methodologies. This implies that the sensor dynamics of the suit incorporate the requisite motion information. In comparison, under these conditions, LSTM and GRU, which both employ a gradient method to calculate the network's weights, do not

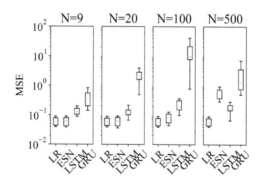

FIGURE 4.5 Mean squared error of different readout modules with different network nodes. The red lines show medians. N indicates the number of network nodes used in ESN, LSTM, and GRU. Note that the vertical axis is a logarithmic axis [12].

achieve sufficient convergence, leading to a higher MSE. This trend is particularly pronounced with the increase in the number of nodes. On the other hand, the ESN achieved a lower MSE due to its exclusive adjustment of the output layer. However, a higher MSE was observed when the number of nodes reached 500.

Figure 4.6 offers an instance of the actual output derived from the readout modules for a swinging back and forth action with a bent arm (R1), where $N=20$. The solid line represents the predicted trajectory from each readout module, and the dotted lines illustrate the actual trajectories. Compared to LR and ESN, LSTM and GRU align with the other trajectories, excluding the sixth joint where the control signal was delivered.

4.4 CONCLUSION AND DISCUSSION

We proposed a novel approach to physical reservoir computing through the implementation of a wearable tactile sensor suit, demonstrating a fabric-based information processing paradigm that utilizes tactile sensor networks as computational resources when worn by an android robot. Despite the limited number of sensors utilized (nine-channel tactile sensors on the left arm), our results demonstrated comparable emulation performance between a LR model of sensor states, which lacks memory capacity, and an ESN, which is a nonlinear recurrent neural network, for the android robot wearing the suit. This implies that fabric-based tactile sensors offer potential as computational tools in motion emulation tasks.

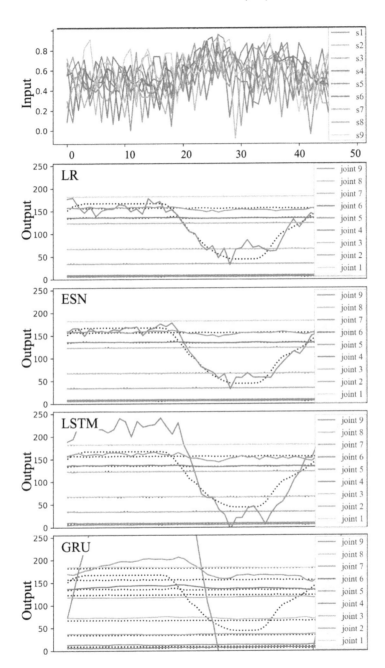

FIGURE 4.6 Performance of four readout modules for motion R1 (swing with bent arm) in $N=20$: Top figure shows normalized time series data of touch sensors. Each line shows sensor value of each touch sensor. Other four figures show performances of linear regression (LR), echo state network (ESN), long short-term memory network (LSTM), and gated recurrent unit network (GRU) [12].

In terms of the MSE, ESN and LR were comparable, whereas the LSTM and GRU networks demonstrated higher MSEs due to insufficient training data and shorter learning times. Given their efficiency, with modifications restricted to the output layer's weights, ESN or LR is likely a better choice for real-time motion emulation.

Past research has argued that ESNs tend to overfit more readily than LSTM or GRU networks as the number of nodes increases [18]. This might account for the increase in MSE observed in ESN with 500 nodes, suggesting that the number of nodes must be carefully optimized depending on the task at hand. ESNs may also not be suitable for predicting systems with hidden states [18]; LSTM or GRU networks might perform better in such cases.

In contrast to gradient methods like LSTM and GRU, fabric-based information processing only learns linear readouts without modifying the clothing structure. Therefore, readouts can be added as needed, opening possibilities for new applications that enhance social interaction. For instance, a conceivable implementation is a speaking function in response to sign language.

Although the present study demonstrated the potential of a fabric-based tactile sensor suit as a computational resource, it does have limitations. Since we only performed monotonous and periodic motions, further testing is needed with more complex memory-demanding trajectories. Similarly, the utilization of a full-body suit equipped with a greater number of tactile sensors warrants investigation to understand the impact of the number of sensors on the learning process.

ACKNOWLEDGMENT

This research was supported by JST CREST (Grant Number JPMJCR18A1 and JPMJCR2014) and JSPS KAKENHI (Grant Number JP20K11915 and JP18H05472), Japan. The results were partially obtained from a project commissioned by NEDO.

REFERENCES

[1] H. Sumioka, T. Minato, and M. Shiomi, "Development of a sensor suit for touch and pre-touch perception toward close human-robot touch interaction," In *RoboTac 2019: New Advances in Tactile Sensation, Perception, and Learning in Robotics: Emerging Materials and Technologies for Manipulation in the IEEE/RSJ International Conference on Intelligent Robots and Systems*, Macaou, China, 2019, p. 3.

[2] M. Shiomi, H. Sumioka, K. Sakai, T. Funayama, and T. Minato, "SŌTO: An android platform with a masculine appearance for social touch interaction," In *The 15th Annual ACM/IEEE International Conference on Human Robot Interaction (HRI2020)*, Cambridge United Kingdom, Mar. 2020, pp. 447–449. doi: 10.1145/3371382.3378283.

[3] G. Tanaka et al., "Recent advances in physical reservoir computing: A review," *Neural Netw.*, vol. 115, pp. 100–123, Jul. 2019, doi: 10.1016/j. neunet.2019.03.005.

[4] K. Nakajima, "Physical reservoir computing—An introductory perspective," *Jpn. J. Appl. Phys.*, vol. 59, no. 6, p. 060501, Jun. 2020, doi: 10.35848/1347-4065/ab8d4f.

[5] H. Sumioka, H. Hauser, and R. Pfeifer, "Computation with mechanically coupled springs for compliant robots," In *Proceedings of the IEEE/RSJ International Conference on Intelligent Robots and Systems*, San Francisco, CA, USA, 2011, pp. 4168–4173.

[6] Q. Zhao, K. Nakajima, H. Sumioka, X. Yu, and R. Pfeifer, "Embodiment enables the spinal engine in quadruped robot locomotion," In *2012 IEEE/RSJ International Conference on Intelligent Robots and Systems*, Vilamoura-Algarve, Portugal, Oct. 2012, pp. 2449–2456. doi: 10.1109/ IROS.2012.6386048.

[7] K. Nakajima, H. Hauser, T. Li, and R. Pfeifer, "Information processing via physical soft body," *Sci. Rep.*, vol. 5, no. 1, Art. no. 1, May 2015, doi: 10.1038/ srep10487.

[8] K. Nakajima, T. Li, H. Hauser, and R. Pfeifer, "Exploiting short-term memory in soft body dynamics as a computational resource," *J. R. Soc. Interface*, vol. 11, no. 100, p. 20140437, Nov. 2014, doi: 10.1098/rsif.2014.0437.

[9] K. Nakajima, H. Hauser, R. Kang, E. Guglielmino, D. G. Caldwell, and R. Pfeifer, "A soft body as a reservoir: Case studies in a dynamic model of octopus-inspired soft robotic arm," *Front. Comput. Neurosci.*, vol. 7, p. 91, Jul. 2013, doi: 10.3389/fncom.2013.00091.

[10] K. Nakajima, H. Hauser, T. Li, and R. Pfeifer, "Exploiting the dynamics of soft materials for machine learning," *Soft Robot.*, vol. 5, no. 3, pp. 339–347, Jun. 2018, doi: 10.1089/soro.2017.0075.

[11] W. Shi, J. Cao, Q. Zhang, Y. Li, and L. Xu, "Edge computing: Vision and challenges," *IEEE Internet Things J.*, vol. 3, no. 5, pp. 637–646, Oct. 2016, doi: 10.1109/JIOT.2016.2579198.

[12] H. Sumioka, K. Nakajima, K. Sakai, T. Minato, and M. Shiomi, "Wearable Tactile Sensor Suit for Natural Body Dynamics Extraction: Case Study on Posture Prediction Based on Physical Reservoir Computing," 2021 IEEE/RSJ International Conference on Intelligent Robots and Systems (IROS), Prague, Czech Republic, pp. 9504–9511, 2021, doi: 10.1109/IROS51168.2021.9636194.

[13] H. Jaeger, "Harnessing nonlinearity: Predicting chaotic systems and saving energy in wireless communication," *Science*, vol. 304, no. 5667, pp. 78–80, Apr. 2004, doi: 10.1126/science.1091277.

[14] S. Hochreiter and J. Schmidhuber, "Long short-term memory," *Neural Comput.*, vol. 9, no. 8, pp. 1735–1780, Nov. 1997, doi: 10.1162/neco.1997.9.8.1735.

[15] K. Cho, B. van Merrienboer, C. Gulcehre, D. Bahdanau, F. Bougares, H. Schwenk, and Y. Bengio, "Learning phrase representations using RNN encoder-decoder for statistical machine translation," *ArXiv14061078 Cs Stat,* Sep. 2014, Accessed: Feb. 23, 2021. [Online]. Available: https://arxiv.org/abs/1406.1078.

[16] M. Lukoševičius and H. Jaeger, "Reservoir computing approaches to recurrent neural network training," *Comput. Sci. Rev.,* vol. 3, no. 3, pp. 127–149, Aug. 2009, doi: 10.1016/j.cosrev.2009.03.005.

[17] D. P. Kingma and J. Ba, "Adam: A method for stochastic optimization," *ArXiv14126980 Cs,* Jan. 2017, Accessed: Feb. 03, 2021. [Online]. Available: https://arxiv.org/abs/1412.6980.

[18] P. R. Vlachas, J. Pathak, B. R. Hunt, T. P. Sapsis, M. Girvan, E. Ott, and P. Koumoutsakos, "Backpropagation algorithms and reservoir computing in recurrent neural networks for the forecasting of complex spatiotemporal dynamics," *Neural Netw.,* vol. 126, pp. 191–217, Jun. 2020, doi: 10.1016/j.neunet.2020.02.016.

CHAPTER 5

SŌTO

An Android Platform for Social Touch Interaction

Masahiro Shiomi, Hidenobu Sumioka,
Kurima Sakai, and Tomo Funayama
Advanced Telecommunications Research
Institute International, Kyoto, Japan

Takashi Minato
RIKEN, Saitama, Japan

5.1 INTRODUCTION

Robotics research is based on the development of useful hardware that has sufficient capabilities for human interaction because well-designed robot hardware enables researchers to investigate various topics. Therefore, many robotics researchers have especially designed robot platforms that match their demands. For example, in the context of supporting seniors, researchers developed several cute and acceptable robot devices [1–3]. Similar to these contexts, researchers have developed several different types of social robots for conversational interactions [4–8].

Researchers who are working on social touch interaction have also developed several specially designed robots [9–12] because existing robots suffer from several restrictions, including limited joint angles for natural touch interaction and limited sensing systems for social touch interaction. These studies focused on whole-body touch interactions such as hugging

DOI: 10.1201/9781003384274-7

because human-like, intimate whole-body interactions with robots provide several positive effects for interacting people.

Appearance plays a significant role in human–human touch interactions. For instance, a past study showed that women generally respond more favorably to being touched than men [13]. Another study focused on touch interactions between the same and opposite genders and discovered that touching by the opposite gender offers distinct benefits [14]. Note that men recipients have reported negative effects from touch interactions of the same gender in nursing contexts [15].

Shifting the lens to human–robot interaction within the scope of social touch, the main focus has been the effects of the gender of the human participants, given that most robots are gender-neutral [15,16]. Although these studies found positive effects in human–robot touch interaction, the perceived gender effects of the appearance of robots have been inadequately studied. Although some past studies involved androids, they largely portrayed feminine characteristics [17–19]. This situation complicates attempts to examine the influence of a robot's perceived gender in social touch within human–robot interaction. Several androids with a masculine appearance do exist; yet they usually feature a face design borrowed from an actual person's face [20]. This complicates the task of avoiding appearance effects in experiments, especially when the source face belongs to a well-known person. Researchers also developed gender-neutral android robots based on non-existent faces [21–23], although they were not specially designed for human–robot touch interaction.

Taking these factors into account, we developed from computer-graphic face designs an android platform with a masculine appearance (Figure 5.1) and named it SŌTO (SOcial TOuch, "創仁" in Japanese). To facilitate social touch interactions with humans, we outfitted SŌTO with fabric-based, capacitance type upper/lower-body touch sensors. In the remainder of this paper, we delve into the specifics of our android platform. Note that this chapter is modified based on our previous work [24], edited to be comprehensive and fit with the context of this book.

5.2 SYSTEM OVERVIEW

5.2.1 Hardware

For the android's development, we embraced the design policy of an existing android, ERICA [8], and worked closely with A-Lab[1] on both

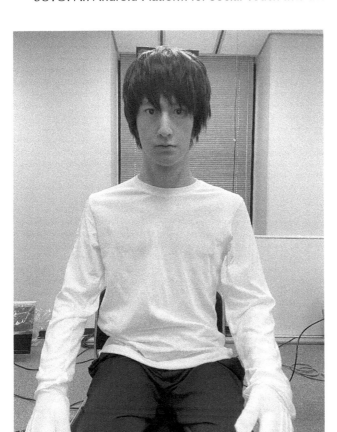

FIGURE 5.1 Photograph of SŌTO: an android platform for social touch interaction with people.

its mechanical and appearance design. Similar to ERICA, the origins of SŌTO's face are found in computer graphics (CG) rather than actual human faces; its face has a symmetrical design [25]. This approach sidesteps any potential copyright or right of refusal to be photographed issues arising from using real human faces. Moreover, using a CG-based face mitigates biases toward appearances. For instance, if we used a Geminoid (featuring Prof. Ishiguro's face [20]) and participants were already aware of his media activities, their perception of the android would undoubtedly be influenced.

In total, SŌTO has 44 degrees of freedoms (DOFs) on its face, torso, and arms. Note that its legs are not yet actuated. We used pneumatic actuators for all the joints except the eyes for smooth and silent movements. The majority of joint control revolves around facial actuation for expressions

FIGURE 5.2   Upper-body touch sensor suit.

and speech. Each eye has three DOFs, offering synchronous control in yaw, pitch, and convergence. The upper and lower eyelids and the inner and outer eyebrows have four DOFs. Another set of four DOFs comes from mouth control: height, width, and the upper and lower corners of the mouth. Last, the tongue and the jaw contribute two DOFs of facial actuation. By using its arm, this android can touch a person's face [26].

We developed a fabric-based touch sensor that senses changes in capacitance. Figure 5.2 illustrates the upper-body touch sensor suit that can detect the location of a touch from a human. We also developed a lower-body touch sensor suit, which has a total of 94 channels that collect a variety of data about touch interactions.

5.2.2 Software

Concerning SŌTO's software aspect, we integrated its control system toward ERICA's software system [8] because their hardware settings are quite similar. Thus, SŌTO possesses a rudimentary and autonomous conversational ability akin to ERICA's and maintains autonomous control over its facial expressions, gaze, and blinking behaviors. We believe that retaining compatibility with an existing android system provides a platform for exploring gender effects in social touch interaction and can capitalize on the variance in hardware systems. In addition, the system autonomously governs speech-related behaviors like lip-syncing, rhythm, and backchannels. For managing dialog and interaction strategies, we employed existing visual tools [27] to facilitate an interaction design process.

Related to our developed touch sensor system, the android can respond to a human touch, for instance, by directing its gaze toward the point of contact. This sensor, which can roughly detect a distance between its sensors and a human body, also enables it to anticipate touching behaviors, e.g., gazing toward a hand before it actually touches the robot [28,29]. In addition, our past study used its upper-body sensor for a data collection to investigate how people physically interact with others and built a function that recognizes such physical interactions by touch sensor outputs. In other words, the android can also detect various kinds of touch interaction with its touch sensor system [30].

5.3 POSSIBLE USE CASES

In this section, we describe possible use cases for SŌTO by considering the existing social touch interaction studies. Note that already we used SŌTO in the context of social touch interaction [28]. Since such opportunities remain limited, we discuss the possibility of using such robots for future work.

Simple but effective cases are comparisons with a feminine-appearance android robot (i.e., ERICA) in the context of social touch interaction. We previously conducted several experiments to understand the effects of social touch interaction [18,19,31,32] and gathered data for modeling touch behaviors based on human touch behaviors [33,34]. Therefore, experimentally using SŌTO with an identical setting as ERICA's might provide interesting knowledge for comparisons.

Other future work will examine the touch interactions between androids that possess both feminine and masculine appearances and humans. This study will explore how an android's appearance influences the perceived feelings surrounding touch interactions with it. We also plan to delve into the combined effects of the perceived genders of androids and the gender of participants through such comparative experiments. Comparisons with robots having machine-like appearances [4,35–37] will enrich our understanding of appearance effects and illuminate compelling design policies for human–robot touch interaction.

Investigating the relationship between touch interaction and the uncanny valley [38] is another future work. Recent studies have focused on its appearance effects [39–41]. A past study also investigated the uncanny valley effects in human perception of forces feedback in a VR environment [42]. However, the effects of touch behaviors on the uncanny valley in human–robot interaction have not been adequately investigated. We believe that human-like natural touch motion and touch characteristics are essential for more acceptable interactions because people may act tentatively around robots that touch unnaturally even though their appearance is quite human-like.

Although we developed SŌTO as an android platform for social touch interaction, of course, it can be employed for conversational interaction without any touch interaction. As described above, since SŌTO's interactive functions resemble ERICA's, investigating masculine appearance effects in conversational interaction is one promising future work that can follow the same settings of past studies that focused on human–robot conversational interaction [43,44].

5.4 CONCLUSION

This paper introduces SŌTO, an android platform developed explicitly for investigating social touch interaction between humans and robots. It outlines the android's hardware, which includes a sensor suit equipped with fabric-based touch sensors, as well as the details of its software configurations. We also discuss possible use cases with our specially designed robot for social touch interaction by referring to current studies related to human–robot interaction. We anticipate that SŌTO will support discerning the differences in human–human and human–robot interactions within the scope of social touch.

ACKNOWLEDGMENT

This research work was supported in part by JST CREST Grant Number JPMJCR18A1, Japan.

NOTE

1 http://www.a-lab-japan.co.jp/en/

REFERENCES

[1] T. Shibata, "An overview of human interactive robots for psychological enrichment," *Proceedings of the IEEE*, vol. 92, no. 11, pp. 1749–1758, 2004.

[2] M. Kanoh, "Babyloid," *Journal of Robotics and Mechatronics*, vol. 26, no. 4, pp. 513–514, 2014.

[3] H. Sumioka, N. Yamato, M. Shiomi, and H. Ishiguro, "A minimal design of a human infant presence: A case study toward interactive doll therapy for older adults with dementia," *Frontiers in Robotics and AI*, vol. 8, p. 633378, 2021.

[4] T. Kanda, H. Ishiguro, T. Ono, M. Imai, and R. Nakatsu, "Development and evaluation of an interactive humanoid robot "Robovie"," In *Proceedings 2002 IEEE International Conference on Robotics and Automation (Cat. No.02CH37292)*, vol. 2, pp. 1848–1855, 2002.

[5] A. van Breemen, X. Yan, and B. Meerbeek, "iCat: An animated user-interface robot with personality," In *Proceedings of the Fourth International Joint Conference on Autonomous Agents and Multiagent Systems*, pp. 143–144, 2005.

[6] G. Metta, G. Sandini, D. Vernon, L. Natale, and F. Nori, "The iCub humanoid robot: An open platform for research in embodied cognition," In *Proceedings of the 8th workshop on performance metrics for intelligent systems*, pp. 50–56, 2008.

[7] J. Kory, and C. Breazeal, "Storytelling with robots: Learning companions for preschool children's language development," In *Robot and Human Interactive Communication, 2014 RO-MAN: The 23rd IEEE International Symposium on*, pp. 643–648, 2014.

[8] D. F. Glas, T. Minato, C. T. Ishi, T. Kawahara, and H. Ishiguro, "Erica: The erato intelligent conversational android," In *Robot and Human Interactive Communication (RO-MAN), 2016 25th IEEE International Symposium on*, pp. 22–29, 2016.

[9] S. Yohanan, and K. E. MacLean, "The role of affective touch in human-robot interaction: Human intent and expectations in touching the haptic creature," *International Journal of Social Robotics*, vol. 4, no. 2, pp. 163–180, 2012.

[10] A. E. Block, and K. J. Kuchenbecker, "Softness, warmth, and responsiveness improve robot hugs," *International Journal of Social Robotics*, vol. 11, no. 1, pp. 49–64, 2019.

[11] M. Shiomi, A. Nakata, M. Kanbara, and N. Hagita, "Robot reciprocation of hugs increases both interacting times and self-disclosures," *International Journal of Social Robotics*, vol. 13, pp. 353–361, 2021.

[12] A. E. Block, H. Seifi, O. Hilliges, R. Gassert, and K. J. Kuchenbecker, "In the arms of a robot: Designing autonomous hugging robots with intra-hug gestures," *ACM Transactions on Human-Robot Interaction*, vol. 12, pp. 1–49, 2022.

[13] D. S. Stier, and J. A. Hall, "Gender differences in touch: An empirical and theoretical review," *Journal of Personality and Social Psychology*, vol. 47, no. 2, p. 440, 1984.

[14] A. S. Ebesu Hubbard, A. A. Tsuji, C. Williams, and V. Seatriz, "Effects of touch on gratuities received in same-gender and cross-gender dyads," *Journal of Applied Social Psychology*, vol. 33, no. 11, pp. 2427–2438, 2003.

[15] M. Shiomi, K. Nakagawa, K. Shinozawa, R. Matsumura, H. Ishiguro, and N. Hagita, "Does a robot's touch encourage human effort?," *International Journal of Social Robotics*, vol. 9, pp. 5–15, 2016.

[16] R. Andreasson, B. Alenljung, E. Billing, and R. Lowe, "Affective touch in human-robot interaction: Conveying emotion to the nao robot," *International Journal of Social Robotics*, vol. 10, no. 4, pp. 473–491, 2018.

[17] M. Shiomi, T. Minato, and H. Ishiguro, "Subtle reaction and response time effects in human-robot touch interaction," In *International Conference on Social Robotics*, pp. 242–251, 2017.

[18] M. Shiomi, K. Shatani, T. Minato, and H. Ishiguro, "How should a robot react before people's touch?: Modeling a pre-touch reaction distance for a robot's face," *IEEE Robotics and Automation Letters*, vol. 3, no. 4, pp. 3773–3780, 2018.

[19] X. Zheng, M. Shiomi, T. Minato, and H. Ishiguro, "What kinds of robot's touch will match expressed emotions?," *IEEE Robotics and Automation Letters*, vol. 5, pp. 127–134, 2019.

[20] D. Sakamoto, and H. Ishiguro, "Geminoid: Remote-controlled android system for studying human presence," *Kansei Engineering International*, vol. 8, no. 1, pp. 3–9, 2009.

[21] H. Ishihara, and M. Asada, "Design of 22-DOF pneumatically actuated upper body for child android 'Affetto'," *Advanced Robotics*, vol. 29, no. 18, pp. 1151–1163, 2015.

[22] Y. Nakata, S. Yagi, S. Yu, Y. Wang, N. Ise, Y. Nakamura, and H. Ishiguro, "Development of 'ibuki'an electrically actuated childlike android with mobility and its potential in the future society," *Robotica*, vol. 40, no. 4, pp. 933–950, 2022.

[23] W. Sato, S. Namba, D. Yang, S. Y. Nishida, C. Ishi, and T. Minato, "An android for emotional interaction: Spatiotemporal validation of its facial expressions," *Frontiers in Psychology*, vol. 12, p. 6521, 2022.

[24] M. Shiomi, H. Sumioka, K. Sakai, T. Funayama, and T. Minato, "SŌTO: An android platform with a masculine appearance for social touch interaction," In *Companion of the 2020 ACM/IEEE International Conference on Human-Robot Interaction*, Cambridge, United Kingdom, pp. 447–449, 2020.

[25] D. I. Perrett, D. M. Burt, I. S. Penton-Voak, K. J. Lee, D. A. Rowland, and R. J. E. Edwards, "Symmetry and human facial attractiveness," *Evolution and Human Behavior*, vol. 20, no. 5, pp. 295–307, 1999.

[26] A. E. Fox-Tierney, K. Sakai, M. Shiomi, T. Minato, and H. Ishiguro, "Sawarimōto: A vision and touch sensor-based method for automatic or tele-operated android-to-human face touch," In *Companion of the 2023 ACM/IEEE International Conference on Human-Robot Interaction*, pp. 481–485, 2023.

[27] D. Glas, S. Satake, T. Kanda, and N. Hagita, "An interaction design framework for social robots," In *Robotics: Science and Systems, University of Southern California*, Los Angeles, CA, USA, June 27–30, p. 89, 2011.

[28] D. A. C. Mejía, H. Sumioka, H. Ishiguro, and M. Shiomi, "Modeling a pre-touch reaction distance around socially touchable upper body parts of a robot," *Applied Sciences*, vol. 11, no. 16, p. 7307, 2021.

[29] D. A. C. Mejía, H. Sumioka, H. Ishiguro, and M. Shiomi, "Evaluating gaze behaviors as pre-touch reactions for virtual agents," *Frontiers in Psychology*, vol. 14, p. 1129677, 2023.

[30] S. Keshmiri, M. Shiomi, H. Sumioka, T. Minato, and H. Ishiguro, "Gentle versus strong touch classification: Preliminary results, challenges, and potentials," *Sensors*, vol. 20, no. 11, p. 3033, 2020.

[31] M. Shiomi, K. Shatani, T. Minato, and H. Ishiguro, "Does a robot's subtle pause in reaction time to people's touch contribute to positive influences?," In *2018 27th IEEE International Symposium on Robot and Human Interactive Communication (RO-MAN)*, pp. 364–369, 2018.

[32] X. Zheng, M. Shiomi, T. Minato, and H. Ishiguro, "How can robots make people feel intimacy through touch?," *Journal of Robotics and Mechatronics*, vol. 32, no. 1, pp. 51–58, 2020.

[33] X. Zheng, M. Shiomi, T. Minato, and H. Ishiguro, "Modeling the timing and duration of grip behavior to express emotions for a social robot," *IEEE Robotics and Automation Letters*, vol. 6, no. 1, pp. 159–166, 2020.

[34] M. Shiomi, X. Zheng, T. Minato, and H. Ishiguro, "Implementation and evaluation of a grip behavior model to express emotions for an android robot," *Frontiers in Robotics and AI*, vol. 8, p. 755150, 2021.

[35] R. Matsumura, M. Shiomi, K. Nakagawa, K. Shinozawa, and T. Miyashita, "A desktop-sized communication robot: "Robovie-mr2"," *Journal of Robotics and Mechatronics*, vol. 28, no. 1, pp. 107–108, 2016.

[36] M. Shiomi, and N. Hagita, "Audio-visual stimuli change not only robot's hug impressions but also its stress-buffering effects," *International Journal of Social Robotics*, vol. 13, pp. 469–476, 2021.

[37] R. Matsumura, and M. Shiomi, "An animation character robot that increases sales," *Applied Sciences*, vol. 12, no. 3, p. 1724, 2022.

[38] M. Mori, K. F. MacDorman, and N. Kageki, "The uncanny valley [from the field]," *IEEE Robotics & Automation Magazine*, vol. 19, no. 2, pp. 98–100, 2012.

[39] K. F. MacDorman, "Subjective ratings of robot video clips for human likeness, familiarity, and eeriness: An exploration of the uncanny valley," In *ICCS/CogSci-2006 Long Symposium: Toward Social Mechanisms of Android Science*, 2006.

[40] A. M. Rosenthal-Von Der Pütten, and N. C. Krämer, "How design characteristics of robots determine evaluation and uncanny valley related responses," *Computers in Human Behavior*, vol. 36, pp. 422–439, 2014.

[41] M. Mara, M. Appel, and T. Gnambs, "Human-like robots and the uncanny valley: A meta-analysis of user responses based on the godspeed scales," *Zeitschrift für Psychologie*, vol. 230, no. 1, p. 33, 2022.

[42] C. C. Berger, M. Gonzalez-Franco, E. Ofek, and K. Hinckley, "The uncanny valley of haptics," *Science Robotics*, vol. 3, no. 17, p. eaar7010, 2018.

[43] T. Uchida, H. Takahashi, M. Ban, J. Shimaya, Y. Yoshikawa, and H. Ishiguro, "A robot counseling system—What kinds of topics do we prefer to disclose to robots?," In *2017 26th IEEE International Symposium on Robot and Human Interactive Communication (RO-MAN)*, pp. 207–212, 2017.

[44] T. Uchida, H. Takahashi, M. Ban, J. Shimaya, T. Minato, K. Ogawa, Y. Yoshikawa, and H. Ishiguro, "Japanese Young Women did not discriminate between robots and humans as listeners for their self-disclosure-pilot study," *Multimodal Technologies and Interaction*, vol. 4, no. 3, p. 35, 2020.

SECTION 3

Modeling Pre-touch Proxemics

CHAPTER 6

Implementing Pre-touch Reaction Distance around Face for a Social Robot

Masahiro Shiomi

Advanced Telecommunications Research Institute International, Kyoto, Japan

Kodai Shatani

Osaka University, Osaka, Japan

Takashi Minato

RIKEN, Saitama, Japan

Hiroshi Ishiguro

Osaka University, Osaka, Japan

6.1 INTRODUCTION

The positive physical and mental effects of haptic interaction have been widely reported in human psychology literature [1–6]. Based on these results, researchers in human–robot interaction have investigated whether

DOI: 10.1201/9781003384274-9

haptic interaction with robots provides such effects. Several studies found positive results and the advantages of using robots with haptic interactions. For example, researchers developed a seal robot for elderly care and reported its mental therapy effects [7]. Another study reported the stress-buffering effects of haptic devices [8]. Touch interaction with robots caused behaviour changes in people, including motivation improvements [9,10], prosocial behaviour encouragement [11,12], and self-disclosure encouragement [13,14]. Moreover, touch interaction design is essential for expressing robot's emotions [15–19]. These studies reported the effectiveness of haptic interaction with social robots and provided knowledge about designing their touch behaviours.

Unfortunately, these studies focused less on the pre-touching situation, although they rigorously investigated the effectiveness of interaction after actual touching. Therefore, it remains unknown how robots should behave before being touched. We believe that pre-touch interaction design is critical for achieving natural touch interaction. In fact, robots need to behave appropriately before any kind of interaction, e.g., distancing behaviours, to appropriately adjust their positioning before conversational interaction.

Related to conversational interaction, E.T. Hall proposed the groundbreaking theory of proxemics [20], which has greatly influenced the design of robot's behaviours. Several robotics researchers have used Hall's theories to design robot's distancing behaviours because adjusting position relationships with interacting persons influenced the social acceptance of robots [21–26]. We investigated the pre-touch proxemics concerning natural behaviour design in before-touch situations.

For this purpose, first, we focused on the pre-touch reaction distance around faces in human–human touch interaction. Although we only investigated the distance around the face, such data will be useful for constructing a foundation for the pre-touch behaviour design of social robots in the context of human–robot touch interaction. To measure the pre-touch reaction distance in human–human touch interaction, we conducted a data collection with pairs of participants. During the experiment, one participant moved his/her hand towards the face of another participant, and we measured the pre-touch reaction distance at which the latter participant wanted the approaching hand to stop. Then, we analysed the collected data and implemented pre-touch reactions based on the data collection results with ERICA (Figure 6.1) [27]. We also investigated the perceived impressions towards ERICA with/without pre-touch reaction in an experiment with human participants. Note that this chapter is modified

FIGURE 6.1   ERICA reacts before being touched.

based on our previous work [28], edited to be comprehensive and fit with the context of this book.

6.2 ANALYSING PRE-TOUCH REACTION DISTANCE

6.2.1 Data Collection Procedure

For the data collection, we employed similar approaches to past studies that measured the personal distance between participants and robots [21–26]. In our data collection, we prepared two roles: touchers and evaluators. After fixing the evaluators' positions and their face directions, the touchers slowly moved their hands towards the faces of the evaluators. When the latter felt discomfort, they explicitly vocalized such feelings and clicked on a computer mouse, and the touchers immediately stopped their hand movements. We then measured the hand's distance from the face and labeled it as their preferred pre-touch reaction distance. We covered various angles around the faces by allowing the touchers to move their hands towards the evaluators from any direction. The evaluators sat on a chair and looked at a mark to maintain their gaze direction during the experiment. The touchers stood around the evaluator at a certain distance. We measured the pre-touch reaction distances from various angles of nine standing positions. The number of touches varied between participant pairs because each pair repeated the data collection as much as possible within the time allotted to them.

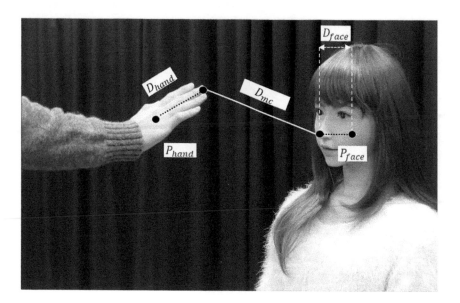

FIGURE 6.2   Distance definitions in our data collection.

6.2.2 Recording Data

We prepared an autonomous tracking system to record the relative positions of the touchers' hands and the evaluators' faces. Figure 6.2 shows distance variables in our data collection. We used two Kinect V2 sensors and calibrated their relative positions for accurate data collection. We also used Kinect V2's library to detect the centre positions of the hands of the touchers (P_{hand} in Figure 6.2) and the head positions of the evaluators (P_{face} in Figure 6.2). We also measured the timing of the mouse clicks by the evaluators to calculate the pre-touch reaction distance between a toucher's hand and an evaluator's face (D_{mc} in Figure 6.2). We measured the size of the hands of the touchers before the experiment and added such information to the calculations and also used the average size of Japanese faces based on a previous study [29] (D_{face} in Figure 6.2, 9 cm for women and 10 cm for men).

6.2.3 Analysis

Forty pairs of participants joined our data collection: ten pairs of female touchers and female evaluators, ten pairs of female touchers and male evaluators, ten pairs of male touchers and female evaluators, and ten pairs of male touchers and male evaluators. The average age of the participants was 21.83, and their *SD* was 1.53. The valid distance data were 11,699, the average number of each toucher was 292.5, and the SD was 81.1. Since each evaluator's preferred pre-touch reaction distance differed and each toucher's approached at a different speed, the number of data for each pair is also different. Therefore, we used the averaged pre-touch reaction distance for our analysis.

The participants' average pre-touch reaction distance was 19.97 cm, and their SD was 8.48. We analysed the effects of genders of the touchers/evaluators and the touch angles. First, we analysed the gender effects by conducting a two-factor ANOVA with the pre-touch reaction distance as a dependent variable, the toucher's gender, and the evaluator's gender. There were no significant differences in the toucher's gender $(F(1, 39)=.012, p=.914, \eta^2=.001)$, the evaluator's gender $(F(1, 39)=.020, p=.888, \eta^2=.001)$, or their interaction $(F(1, 39)=3.844, p=.058, \eta^2=.096)$. Although past studies reported that women preferred shorter personal distances than men [30,31], our data collection found no significant effects on the gender factor in the context of the pre-touch reaction distance.

Next, we investigated the angle effects of the pre-touch reaction distance. Because the gender effects are not significant in this data collection, we used the averaged pre-touch reaction distance by integrating both genders in this analysis. We investigated the angle effects, by separately investigating the left/right side touches and above/below side touches to avoid an excessive number of combinations in the statistical analysis. We conducted a paired t-test about the left/right side touches and found no significant differences $(t(39)=0.188, p=.851, d=.003)$.

Next, we conducted a paired *t*-test for the above/below side touches and identified no significant differences $(t(39)=1.135, p=.263, d=.018)$. Therefore, this data collection showed no significant effects of the angle factor in the context of the pre-touch reaction distance.

Based on the above analysis, we used the average pre-touch reaction distance of all the gathered data without considering the gender/angle differences, i.e., 20 cm as the pre-touch reaction distance.

6.3 EXPERIMENT

6.3.1 Hypothesis and Predictions

We developed a pre-touch reaction model based on the observed pre-touch reaction data to identify an appropriate pre-touch reaction distance. In the modelling process, among the more minor effects of the gender and angle factors, we decided to use 20 cm as a threshold of reaction timing because it is the average pre-touch reaction distance in the data collection. If we appropriately model the pre-touch reaction distance, a robot that reacts towards touch behaviour at 20 cm around its face will be perceived as more human-like and natural than a robot that reacts after being touched or at an intimate distance to a potential touch. Based on these considerations, we made the following prediction:

Prediction: If a robot reacts at 20 cm to a touching behaviour, it will be perceived as *more human-like* and *natural* than a robot that reacts after being touched or at 45 cm.

6.3.2 Robot System

We again used ERICA [27] and a Kinect V2 sensor to detect the positions of the participants' hands. We employed a PCL library [32] to make a clustering 3D object and FLANN [33] to measure the distance between ERICA's face and the participants' hands (D_{reaction}). Our developed system calculated the distance in less than 100 msec on average.

6.3.3 Conditions

We employed a within-participant design for our experiment that has three different conditions. Each participant interacted with ERICA three times. The order of the three conditions was counterbalanced to avoid order effects. The reaction behaviour is looking at the participants' faces in every condition.

Touch condition: D_{reaction} is 0 cm. ERICA reacts after the participants actually touched its nose. Only in this condition, an operator controlled the robot's reaction behaviour to accurately react to the timing of being touched.

Proposed condition: D_{reaction} is 20 cm. Thus, the robot autonomously reacts to being touched by participants when D_{reaction} is less than 20 cm.

Intimate-distance condition: D_{reaction} is 45 cm. Thus, the robot autonomously reacts to being touched by participants when D_{reaction} is less than 45 cm.

6.3.4 Participants

This experiment had 30 participants (15 women and 15 men) whose average ages were 23.0 and *SD* 2.58.

6.3.5 Procedure

Before the experiment, the participants were given a brief description of its purpose and procedure. This research was approved by our institution's ethics committee for studies involving human participants. Written, informed consent was obtained from each of them.

First, we explained the robot's pre-touch reaction behaviour towards a touching behaviour. We asked them to move their hand at a slow and constant speed to avoid speed effects that resemble general touching behaviours. In the experiment, participants moved their hands 18 times (three different positions: left-side, front, and right-side, three different angles: above, front, and below, and two hands: left and right) during the experiment. We counterbalanced each condition's order of positions, angles, and hands. After each condition, the participants completed questionnaires.

6.3.6 Measurements

We prepared two questionnaire items to investigate the feeling of human-likeness ("I think that the robot's reaction distance is human-like") and naturalness ("I think that the robot's reaction distance is natural") of the robot's pre-touch reaction behaviour. The items were assessed on a 1-to-7 response format, where 1 indicates the most negative and 7 indicates the most positive.

6.4 RESULTS

Figure 6.3 shows the results of the human-likeness of the reaction distance. We conducted a two-factor mixed ANOVA for the gender and distance factors, and its results showed significant differences in the distance factor ($F(2, 56)=25.783$, $p<0.001$, $partial\ \eta^2=0.479$), but no significant differences in the gender factor ($F(1, 28)=2.240$, $p=0.146$, $partial\ \eta^2=0.074$) or the interaction effect ($F(2, 56)=0.680$, $p=0.511$, $partial\ \eta^2=0.024$). Multiple comparisons with the Bonferroni method revealed a significant difference

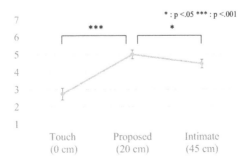

FIGURE 6.3   Questionnaire results of human-likeness of reaction distance: Only significant differences compared to proposed conditions are shown.

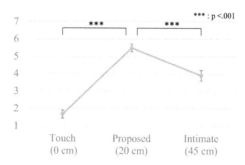

FIGURE 6.4   Questionnaire results of naturalness of reaction distance: Only significant differences compared to proposed conditions are shown.

for the distance factors: proposed > intimate ($p=0.040$), proposed > touch ($p < 0.001$), and intimate > touch ($p < 0.001$).

Figure 6.4 shows the questionnaire results about the naturalness of the reaction distance. We conducted a two-factor mixed ANOVA for the gender and distance factors, and the results showed significant differences in the distance factor ($F(2, 56)=71.493$, $p < 0.001$, $partial\ \eta^2=0.719$). We found a significant trend in the gender factor ($F(1, 28)=3.891$, $p=0.058$, $partial\ \eta^2=0.122$), but no significant difference in the interaction effect ($F(2, 56)=0.579$, $p=0.564$, $partial\ \eta^2=0.020$).

Multiple comparisons with the Bonferroni method revealed a significant difference for the distance factors: proposed > intimate ($p < 0.001$), proposed > touch ($p < 0.001$), and intimate > touch ($p < 0.001$).

Therefore, the participants perceived more human-likeness and naturalness for the reaction distance in the proposed condition than in the alternative conditions, supporting our prediction.

6.5 DISCUSSION

The experiment results suggested several implementations. First, the data collection results showed that gender effects are not significant in the context of pre-touch reaction distance. Past studies related to personal distance reported that evaluators' genders are insignificant when the evaluators are approached by participants [34–38]. On the other hand, some studies reported that the approacher's gender causes significant effects in identical situations [36–38]. These differences between personal and pre-touch reaction distance indicate that people have different perceptions of approaching and touching persons.

The experiment results show the importance of a robot's pre-touch reaction behaviour for natural human–robot touch interaction. The participants perceived more human-likeness and naturalness when the robot reacted to their touching behaviours at a modelled distance. We expected the robot's reaction to fit the participants' assumption because its appearance is human-like, a result that indicates that the robot's appearance is one essential factor in determining an appropriate pre-touch reaction distance. For example, in a similar experiment, if we use a robot with a masculine-like appearance [39], people may prefer a 20-cm pre-touch reaction distance because the data collection results did not show any significant effects of gender factors. However, if we use robots with a different appearance, such as mechanical-robot-like [40–42], creature-like [43], baby-like [44] or animal-like [7], the preferred pre-touch reaction distance may be different. Different appearances and environments (e.g., virtual environments) might influence the appropriate reaction time, behaviors, and distances [45–49]. Lifestyle changes, e.g., the COVID-19 pandemic, also affect the preferred distance with others [50–52].

6.6 CONCLUSION

Pre-touch reaction behaviour design is essential for natural human–robot touch interaction, particularly for a robot with a human-like appearance, because people will assume human-like reactions towards it. Therefore, we conducted a data collection to investigate the appropriate pre-touch reaction distance by observing human–human touch interactions. Based on observations of more than 10,000 touch-interaction data from 40 pairs of participants, we identified the average pre-touch reaction distance to be about 20 cm.

We implemented a pre-touch reaction behaviour towards an android robot that has a feminine-like appearance and experimentally evaluated the implemented behaviours with 30 participants. We prepared three pre-touch reaction distances: 0 cm, which is literally a being touched distance, 20 cm, which is an observed pre-touch reaction distance from the data collection, and 45 cm, which is an intimate distance in conversation situations. The experiment results showed the advantages of implementing a pre-touch reaction behaviour compared to the alternative conditions.

ACKNOWLEDGEMENT

This work was partially supported by JST CREST Grant, Number JPMJCR18A1, Japan.

REFERENCES

[1] K. M. Grewen, B. J. Anderson, S. S. Girdler, and K. C. Light, "Warm partner contact is related to lower cardiovascular reactivity," *Behavioral Medicine*, vol. 29, no. 3, pp. 123–130, 2003.

[2] S. Cohen, D. Janicki-Deverts, R. B. Turner, and W. J. Doyle, "Does hugging provide stress-buffering social support? A study of susceptibility to upper respiratory infection and illness," *Psychological Science*, vol. 26, no. 2, pp. 135–147, 2015.

[3] B. K. Jakubiak, and B. C. Feeney, "Keep in touch: The effects of imagined touch support on stress and exploration," *Journal of Experimental Social Psychology*, vol. 65, pp. 59–67, 2016.

[4] A. Gallace, and C. Spence, "The science of interpersonal touch: An overview," *Neuroscience & Biobehavioral Reviews*, vol. 34, no. 2, pp. 246–259, 2010.

[5] K. C. Light, K. M. Grewen, and J. A. Amico, "More frequent partner hugs and higher oxytocin levels are linked to lower blood pressure and heart rate in premenopausal women," *Biological Psychology*, vol. 69, no. 1, pp. 5–21, 2005.

[6] T. Field, "Touch for socioemotional and physical well-being: A review," *Developmental Review*, vol. 30, no. 4, pp. 367–383, 2010.

[7] R. Yu, E. Hui, J. Lee, D. Poon, A. Ng, K. Sit, K. Ip, F. Yeung, M. Wong, and T. Shibata, "Use of a therapeutic, socially assistive pet robot (PARO) in improving mood and stimulating social interaction and communication for people with dementia: Study protocol for a randomized controlled trial," *JMIR Research Protocols, vol.* 4, no. 2, 2015.

[8] H. Sumioka, A. Nakae, R. Kanai, and H. Ishiguro, "Huggable communication medium decreases cortisol levels," *Scientific Reports*, vol. 3, p. 3034, 2013.

[9] M. Shiomi, K. Nakagawa, K. Shinozawa, R. Matsumura, H. Ishiguro, and N. Hagita, "Does a robot's touch encourage human effort?," *International Journal of Social Robotics*, vol. 9, pp. 5–15, 2016.

[10] K. Higashino, M. Kimoto, T. Iio, K. Shimohara, and M. Shiomi, "Tactile stimulus is essential to increase motivation for touch interaction in virtual environment," *Advanced Robotics*, vol. 35, no. 17, pp. 1043–1053, 2021.

[11] M. Shiomi, A. Nakata, M. Kanbara, and N. Hagita, "A hug from a robot encourages prosocial behavior," In *2017 26th IEEE International Symposium on Robot and Human Interactive Communication (RO-MAN)*, Lisbon, Portugal, 2017.

[12] C. Bevan, and D. Stanton Fraser, "Shaking hands and cooperation in tele-present human-robot negotiation," In *Proceedings of the Tenth Annual ACM/IEEE International Conference on Human-Robot Interaction*, New York, NY, United States, pp. 247–254, 2015.

[13] M. Shiomi, A. Nakata, M. Kanbara, and N. Hagita, "Robot reciprocation of hugs increases both interacting times and self-disclosures," *International Journal of Social Robotics*, vol. 13, pp. 1–9, 2020.

[14] M. Shiomi, and N. Hagita, "Audio-visual stimuli change not only robot's hug impressions but also its stress-buffering effects," *International Journal of Social Robotics*, vol. 13, pp. 1–8, 2019.

[15] X. Zheng, M. Shiomi, T. Minato, and H. Ishiguro, "What kinds of robot's touch will match expressed emotions?," *IEEE Robotics and Automation Letters*, vol. 5, pp. 127–134, 2019.

[16] X. Zheng, M. Shiomi, T. Minato, and H. Ishiguro, "How can robots make people feel intimacy through touch?," *Journal of Robotics and Mechatronics*, vol. 32, no. 1, pp. 51–58, 2019.

[17] X. Zheng, M. Shiomi, T. Minato, and H. Ishiguro, "Modeling the timing and duration of grip behavior to express emotions for a social robot," *IEEE Robotics and Automation Letters*, vol. 6, no. 1, pp. 159–166, 2020.

[18] M. Shiomi, X. Zheng, T. Minato, and H. Ishiguro, "Implementation and evaluation of a grip behavior model to express emotions for an android robot," *Frontiers in Robotics and AI*, vol. 8, p. 755150, 2021.

[19] M. Teyssier, G. Bailly, C. Pelachaud, and E. Lecolinet, "Conveying emotions through device-initiated touch," *IEEE Transactions on Affective Computing*, 2020.

[20] E. T. Hall, *The Hidden Dimension. Knopf Doubleday Publishing Group*. 1966.

[21] L. Takayama, and C. Pantofaru, "Influences on proxemic behaviors in human-robot interaction," In *2009 IEEE/RSJ International Conference on Intelligent Robots and Systems*, St. Louis, MO, United States, pp. 5495–5502, 2009.

[22] J. Mumm, and B. Mutlu, "Human-robot proxemics: Physical and psychological distancing in human-robot interaction," In *Proceedings of the 6th International Conference on Human-Robot Interaction*, Lausanne, Switzerland, pp. 331–338, 2011.

[23] S. Rossi, M. Staffa, L. Bove, R. Capasso, and G. Ercolano, "User's personality and activity influence on HRI comfortable distances," In *International Conference on Social Robotics*, Tsukuba, Japan, pp. 167–177, 2017.

[24] Hiroi, Yutaka, and Akinori Ito, "Influence of the size factor of a mobile robot moving toward a human on subjective acceptable distance," In Mobile Robots: Current Trends, Zoran Gacovski (Ed.), pp. 177–190, 2011.

[25] Y. Kim, S. S. Kwak, and M.-S. Kim, "Am I acceptable to you? Effect of a robot's verbal language forms on people's social distance from robots," *Computers in Human Behavior*, vol. 29, no. 3, pp. 1091–1101, 2013.

[26] M. Obaid, E. B. Sandoval, J. Złotowski, E. Moltchanova, C. A. Basedow, and C. Bartneck, "Stop! That is close enough. How body postures influence human-robot proximity," In *2016 25th IEEE International Symposium on Robot and Human Interactive Communication (RO-MAN)*, New York, pp. 354–361, 2016.

[27] D. F. Glas, T. Minato, C. T. Ishi, T. Kawahara, and H. Ishiguro, "Erica: The erato intelligent conversational android," In *2016 25th IEEE International Symposium on Robot and Human Interactive Communication (RO-MAN)*, New York, pp. 22–29, 2016.

[28] M. Shiomi, K. Shatani, T. Minato, and H. Ishiguro, "How should a robot react before people's touch?: Modeling a pre-touch reaction distance for a robot's face," *IEEE Robotics and Automation Letters*, vol. 3, no. 4, pp. 3773–3780, 2018.

[29] K. Makiko, and M. Mochimaru, "Japanese head size database 2001 (in Japanese)," *AIST, H16PRO-212*, 2008.

[30] M. Baldassare, and S. Feller, "Cultural variations in personal space," *Ethos*, vol. 3, no. 4, pp. 481–503, 1975.

[31] S. Heshka, and Y. Nelson, "Interpersonal speaking distance as a function of age, sex, and relationship," *Sociometry*, vol. 35, pp. 491–498, 1972.

[32] R. B. Rusu, and S. Cousins, "3d is here: Point cloud library (pcl)," In *2011 IEEE International Conference on Robotics and Automation (ICRA)*, Shanghai, China, pp. 1–4, 2011.

[33] M. Muja, and D. G. Lowe, "Fast approximate nearest neighbors with automatic algorithm configuration,"In *Proceedings of the Fourth International Conference on Computer Vision Theory and Applications*, Lisboa, Portugal, February 5–8, 2009.

[34] E. A. Bauer, "Personal space: A study of blacks and whites," *Sociometry*, pp. 402–408, 1973.

[35] M. J. White, "Interpersonal distance as affected by room size, status, and sex," *The Journal of Social Psychology*, vol. 95, no. 2, pp. 241–249, 1975.

[36] B. A. Barrios, L. C. Corbitt, J. P. Estes, and J. S. Topping, "Effect of a social stigma on interpersonal distance," *The Psychological Record*, vol. 26, no. 3, pp. 343–348, 1976.

[37] M. A. Wittig, and P. Skolnick, "Status versus warmth as determinants of sex differences in personal space," *Sex Roles*, vol. 4, no. 4, pp. 493–503, 1978.

[38] G. T. Long, J. W. Selby, and L. G. Calhoun, "Effects of situational stress and sex on interpersonal distance preference," *The Journal of Psychology*, vol. 105, no. 2, pp. 231–237, 1980.

[39] M. Shiomi, H. Sumioka, K. Sakai, T. Funayama, and T. Minato, "SŌTO: An android platform with a masculine appearance for social touch interaction," In *Companion of the 2020 ACM/IEEE International Conference on Human-Robot Interaction*, Cambridge, UK, pp. 447–449, 2020.

[40] T. Kanda, H. Ishiguro, T. Ono, M. Imai, and R. Nakatsu, "Development and evaluation of an interactive humanoid robot "Robovie"," In *Proceedings of the 2002 IEEE International Conference on Robotics and Automation (Cat. No.02CH37292)*, Washington, DC, vol. 2, pp. 1848–1855, 2002.

[41] R. Matsumura, M. Shiomi, K. Nakagawa, K. Shinozawa, and T. Miyashita, "A desktop-sized communication robot: "robovie-mr2"," *Journal of Robotics and Mechatronics*, vol. 28, no. 1, pp. 107–108, 2016.

[42] R. Matsumura, and M. Shiomi, "An Animation Character Robot That Increases Sales," *Applied Sciences*, vol. 12, no. 3, p. 1724, 2022.

[43] N. Yoshida, S. Yonemura, M. Emoto, K. Kawai, N. Numaguchi, H. Nakazato, S. Otsubo, M. Takada, and K. Hayashi, "Production of character animation in a home robot: A case study of lovot," *International Journal of Social Robotics*, vol. 14, no. 1, pp. 39–54, 2022.

[44] H. Sumioka, N. Yamato, M. Shiomi, and H. Ishiguro, "A minimal design of a human infant presence: A case study toward interactive doll therapy for older adults with dementia," *Frontiers in Robotics and AI*, vol. 8, p. 633378, 2021.

[45] A. Kubota, M. Kimoto, T. Iio, K. Shimohara, and M. Shiomi, "From when to when: Evaluating naturalness of reaction time via viewing turn around behaviors," *Applied Sciences*, vol. 11, no. 23, p. 11424, 2021.

[46] D. A. C. Mejía, H. Sumioka, H. Ishiguro, and M. Shiomi, "Modeling a pre-touch reaction distance around socially touchable upper body parts of a robot," *Applied Sciences*, vol. 11, no. 16, p. 7307, 2021.

[47] D. A. C. Mejía, A. Saito, M. Kimoto, T. Iio, K. Shimohara, H. Sumioka, H. Ishiguro, and M. Shiomi, "Modeling of pre-touch reaction distance for faces in a virtual environment," *Journal of Information Processing*, vol. 29, pp. 657–666, 2021.

[48] M. Kimoto, Y. Otsuka, M. Imai, and M. Shiomi, "Effects of appearance and gender on pre-touch proxemics in virtual reality," *Frontiers in Psychology*, vol. 14, p. 1195059, 2023.

[49] D. A. C. Mejía, H. Sumioka, H. Ishiguro, and M. Shiomi, "Evaluating gaze behaviors as pre-touch reactions for virtual agents," *Frontiers in Psychology*, vol. 14, p. 1129677, 2023.

[50] V. Mehta, "The new proxemics: COVID-19, social distancing, and sociable space," *Journal of Urban Design*, vol. 25, no. 6, pp. 669–674, 2020.

[51] R. Welsch, M. Wessels, C. Bernhard, S. Thönes, and C. von Castell, "Physical distancing and the perception of interpersonal distance in the COVID-19 crisis," *Scientific Reports*, vol. 11, no. 1, pp. 1–9, 2021.

[52] M. Shiomi, A. Kubota, M. Kimoto, T. Iio, and K. Shimohara, "Stay away from me: Coughing increases social distance even in a virtual environment," *PLoS One*, vol. 17, no. 12, p. e0279717, 2022.

CHAPTER 7

Implementing Pre-touch Reaction Distance around Upper Body Parts for a Social Robot

Dario Alfonso Cuello Mejía, Hidenobu Sumioka, Hiroshi Ishiguro, and Masahiro Shiomi

Advanced Telecommunications Research Institute International, Kyoto, Japan

Osaka University, Osaka, Japan

7.1 INTRODUCTION

Touch interaction with other people provides various positive effects [1–6], and in this context, using social robots as a partner is a promising approach to compensate for the lack of human–human touch interaction, especially since it provides several positive effects [7–15]. Even though the effects of touch interaction with such social robots might fail to fulfill the needs of interaction with actual people, perhaps people's negative situations can be mitigated.

As described above, many research works have described the effectiveness of touch effects in human–robot interaction (HRI). However, these studies mainly focused on after-touch situations; dealing with pre-touch situations has received less focus. A few past research works concentrated

DOI: 10.1201/9781003384274-10

on a robot's pre-touch reaction distance [16] and for a virtual agent [17], both of which focused on faces. Knowledge about pre-touch interactions remains limited. Based on human science literature, since an upper body part (i.e., the shoulders or arms, including the elbows and the hands) is mainly used in touch interaction between others [18], pre-touch interaction must address such touchable upper body parts for natural touch interaction.

We regard pre-touch interaction research as an extension of proxemics in touch interaction contexts. People generally maintain a certain distance from others, and in conversation interaction contexts, such knowledge is called a personal distance [19]. Many developers of social robots have exploited insights from proxemics and borrowed them as the foundation of their positions when interacting with people [20–33]. Inspired by proxemics, our ultimate goal is to establish its basis in the context of pre-touch interaction as pre-touch proxemics. Such basic knowledge is important for social robots that physically interact with people.

We first gathered people's pre-touch reaction data. Each data collection trial involved two participants whose touch target parts were a shoulder, an elbow, and a hand because previous studies reported that these body parts are acceptable for being touched [18]. For obvious issues (including sensitivity to touch interaction), we eliminated the lower body from our target body parts. Next, we conducted a statistical analysis of the collected data and implemented a pre-touch reaction model in a masculine appearance android called SŌTO [29] (Figure 7.1). Thus, this study aims to identify the minimum comfortable distance in human–human touch interaction around the upper body. Note that this chapter is modified based on our previous work [30], edited to be comprehensive and fit with the context of this book.

FIGURE 7.1   SŌTO's pre-touch reaction towards shoulder approach.

7.2 PRE-TOUCH REACTION DISTANCE IN HUMAN INTERACTION

7.2.1 Data Collection Procedure

For the data collection, we defined two roles for the participants: a toucher who makes physical contact with an evaluator who is touched. Similar to past proxemics studies that modeled pre-touch [16,17] and personal distances [23–28], we gathered self-reported preferred pre-touch reaction distance from the evaluators who sat in chairs. The touchers stood around 1.0 m in front of the evaluators. We fixed the latter's positions, including their face direction, and asked them to constantly direct their gaze to the front. A toucher slowly extended his/her hand toward a body part of the evaluator; when the evaluator felt threatened and wanted the hand to stop, he/she generated an audible signal by clicking a button. The toucher immediately stopped his/her hand when hearing the signal and returned to his/her initial position, and the data collection continued with the same procedures. We asked the participants to move as naturally as possible and use their dominant hands. Each trial's duration was two minutes for each body part: a shoulder, an elbow, and a hand. Each procedure for the evaluator's body parts was repeated for two cases: a front-right approach from the evaluator's right and a front-left approach from his/her left. The order of the approaching sides (front-right or front-left) was counterbalanced between participants. Once the procedure was done, the participants changed roles and repeated the data collection.

7.2.2 Recording Data

We used two OptiTrack systems (Acuity Inc.) as a motion capture system to automatically track the positions of the body parts of both the touchers and the evaluators. We placed four markers on the toucher: shoulder, elbow, palm of the hand (anterior), and back of the hand (posterior). We placed three markers on the evaluator: shoulder, elbow, and hand. We also obtained the timing of mouse clicks by the evaluators with which we calculated the pre-touch reaction distance between a toucher's hand and any body part of an evaluator.

7.2.3 Analysis

Sixteen pairs of Japanese participants (16 men and 16 women; eight man–woman pairs, four man–man pairs, and four woman–woman pairs) joined our data collection. All of them met for the first time in the experiment, i.e., they were strangers. Their average ages and *SD* were 36.39 and 11.63, respectively. The data collection gathered 6,593 values of distance data. On average, the touchers' hands approached each evaluator's body part 34 times within two minutes. The total number of touching data was different between participants because of varied pre-touch reaction distances, approaching speeds, etc. Therefore, we used the averaged pre-touch reaction distance for each evaluator (i.e., 32) in the analysis instead of all the raw data. Figure 7.2 shows part of the gathered data around each body part. Note that these positions were relatively transformed based on the 3D model body parts for visualization purposes. The obtained average pre-touch reaction distance for all the participants was 23.46 cm and the SE was 2.1.

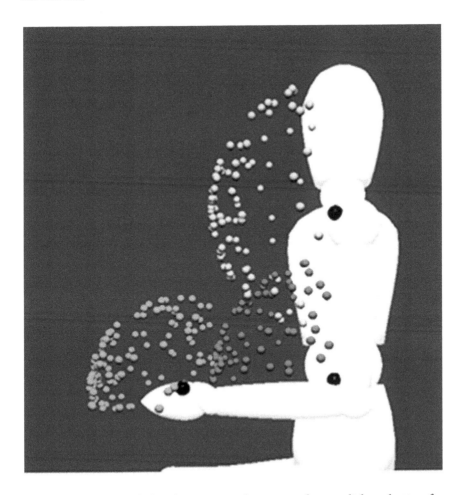

FIGURE 7.2 Toucher's hand positions when an evaluator clicks a button for upper body parts.

TABLE 7.1   Average (*SE*) of Minimum Comfortable Distance in Centimeters

| | Gender Combination | Angle | | |
	Different	Same	Right	Left
Shoulder	25.9 (2.6)	23.4 (2.6)	23.6 (1.8)	25.7 (2.0)
Elbow	24.1 (2.5)	23.6 (2.5)	23.0 (1,6)	24.7 (2.1)
Hand	21.3 (2.3)	21.0 (2.3)	20.6 (1.5)	21.7 (1.8)

First, we analyzed the collected data from three factors: parts (shoulder, elbow, and hand), angle (left and right), and gender combination (same or different) (Table 7.1). We used a gender combination factor because a past study suggested that pre-touch reaction distance increases with the opposite gender without providing strict evidence [16]. We conducted a three-factor mixed ANOVA and only identified a significant main effect in the part factor ($F(2, 60) = 13.905$, $p < 0.001$, partial $\eta^2 = 0.317$). Multiple comparisons with the Bonferroni method revealed a significant difference for the distance factors: shoulder>hand ($p = 0.001$) and elbow>hand ($p = 0.001$). There was no significance between the shoulder and the elbow ($p = 0.591$).

Note that no other factors showed significant effects in all the combinations. Thus, in this data collection, only the body parts significantly affected the pre-touch reaction distance. In other words, the angle and the gender combinations between the touchers and the evaluators did not significantly affect the distance.

Next, we analyzed how the approach speed influenced the data collection, given that the interaction movement differed from person to person. We investigated this effect by investigating the relationship between the speed of the toucher's hand and the minimum comfortable distance and obtained a weak positive correlation ($r = 0.342$, $p < 0.001$). A faster movement might cause more nervousness in the evaluator who reacts more quickly. In our setup, the defined pre-touch situation showed a weak positive correlation for the obtained minimum comfortable distance.

Finally, we analyzed whether the minimum comfortable distance changed during the data collection. Since the participants were involved in many touch interactions, we wanted to determine whether their perception influenced the measured distance. We investigated such an acclimation factor by separating the dataset into two classes: first and second halves, and we used the part factor because we found significant differences in the above analysis. Thus, we again conducted a two-factor (part and acclimation) ANOVA and identified significant main effects in the part factor ($F(2, 62) = 14.216$, $p < 0.001$, partial $\eta^2 = 0.314$) and in the acclimation factor ($F(1, 31) = 18.741$, $p < 0.001$, partial $\eta^2 = 0.377$). We did not find any significance in the interaction effect ($F(2, 62) = 0.591$, $p = 0.557$, partial $\eta^2 = 0.019$). Multiple comparisons with the Bonferroni method revealed a significant difference for the distance factors: shoulder > hand ($p < 0.001$) and elbow > hand ($p < 0.001$). There was no significance between the shoulder and the elbow ($p = 0.653$).

Based on analysis of the pre-touch reaction distance around the upper body parts, we confirmed that the minimum comfortable distance around the hands is smaller than the same distance around the shoulders and elbows. Our results also showed that the gender and angle factors did not show a significant effect and exhibited a similar phenomenon with a past study on pre-touch reaction distance around the face [16]. The parts factor showed a significant difference between the hand and the shoulder/elbow distances. The movement speed showed a weak impact on the minimum comfortable distance; however, the acclimation effect showed a significant difference.

7.3 ROBOT REACTION EXPERIMENT

7.3.1 Implementation

We used an android robot called SŌTO with a masculine appearance [29] and an OptiTrack system that resembles our data collection to capture the body parts' positions in real time. Based on the data collection results, SŌTO reacted to the interaction when a hand approached the threshold distance by using the markers' position information.

In the implementation of the pre-touch reactions for the robot, we prepared three boundaries for the shoulder, elbow, and the hand and another for the face based on a past study [16] to achieve pre-touch reaction behaviors for typical upper body parts. The boundary thresholds are defined as T_{face}, $T_{shoulder}$, T_{elbow}, and T_{hand}, based on the positions of each body part of the robot: P_{Rface}, $P_{Rshoulder}$, P_{Relbow}, and P_{Rhand}. The distances between the participant's hand (Figure 7.3, P_{Phand}) and each body part of the robot were calculated by OptiTrack system's outputs (D_{face}, $D_{shoulder}$, D_{elbow}, and D_{hand}). When this distance was equal to or less than a threshold (e.g., $D_{elbow} \leq T_{elbow}$), the robot reacted and looked at the person using the marker on the participant's head as a reference. In this implementation, we simply employed the average values as the pre-touch reaction distance threshold for defining

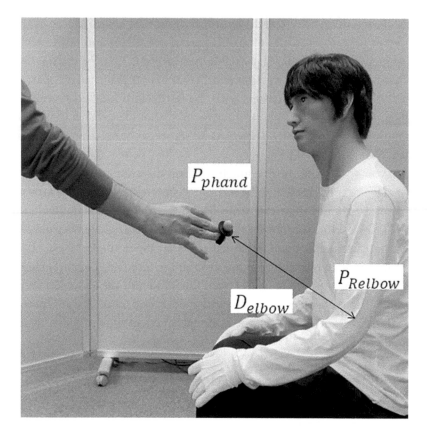

FIGURE 7.3 Reaction to a potential touch to elbow implementing the obtained model.

the model based on these analyses (shoulder = 24.8 cm, elbow = 24.1 cm, and hand = 21.5 cm). Note that the system's thresholds can be changed by considering the angle and the gender of the interacting partners.

7.3.2 Testing

Finally, we tested our developed system. The touchers wore a marker on two fingers of their dominant hand and their head and stood around 1.0 m from the robot's positions. They slowly moved their hands while approaching the robot's body part. The robot was designed to react by looking at the toucher's face, a step that confirms that it autonomously reacted to the pre-touch behaviors. We tested this result with multiple different touchers.

7.4 DISCUSSION

Based on past knowledge and observations, this study implemented a pre-touch reaction distance threshold for the face, shoulders, elbows, and hands. This distance was significantly shorter for the hand than the elbow and shoulder distances. However, these distance thresholds are based on specific body parts. Discussing how to deal with pre-touching for other socially acceptable body parts might be enlightening. We must consider the sensitivity of other body parts because a past study reported that touching such areas is generally unacceptable except by a familiar person [18]. Investigating the relationship between such sensitivity and pre-touch reaction distance is an interesting future work.

Concerning the gender effect, in human science literature, women are more receptive to being touched or engaging in close-distance interaction than men. They are also more willing to accept a same-gender touch than an opposite-gender touch [31–33]. Even though our data collection analysis did not show significant differences, the gathered data may be useful for studying the gender effects of pre-touch distance around body parts.

Cultural differences are another possible factor that might have influenced the data collection results for pre-touch reaction distances. In fact, touch interaction is quite different due to cultural effects [34,35] and situations [36,37]. This study only included Japanese participants in a laboratory setting. Our findings might fuel comparisons of cultural differences in the context of pre-touch interactions.

Participant characteristics such as personality, culture, and age affect the pre-touch reaction distance. In addition, we did not investigate the pre-touch reaction distance around the lower body due to difficulties observing human–human touch interaction. Although building whole-body pre-touch reaction distance knowledge is critical, such data collection is complicated by several issues, including ethical and privacy problems. Even if we used robots or virtual agents as touchers instead of human participants, data collection will inevitably be problematic.

Since our model is based on participants who were basically strangers, our study would undoubtedly change with people who already shared some kind of relationship. Investigating such differences in human–human pre-touch interaction will be interesting for future evaluations of the same effect in HRI.

Moreover, similar to the pre-touch reaction distance around a face, a robot's appearances (e.g., a feminine persona [38], mechanical-robot-like [38–41], and baby-like [42]) will influence the pre-touch reaction distance around body parts. Another factor is changes in lifestyle, i.e., different distance behaviors due to the COVID-19 pandemic [43–45].

Our studies identified a more complex phenomenon than past studies. Our results suggest that robots might require a different strategy before touching and in touch interaction situations due to the appearances and genders of the people with whom it is interacting.

7.5 CONCLUSION

Before-touch reaction behavior is a fundamental approach for improving communication and providing the android with sufficient tools for performing efficient interactions. We evaluated a before-touch situation in human–human touch interactions to define a pre-touch reaction distance and implemented it using a masculine appearance android. We modeled this distance around a (socially acceptable) touchable upper body part, i.e., shoulders, elbows, and hands, based on human–human interaction observations that considered gender, right or left side approach, and the speed of the touch interaction. The pre-touch reaction distances for the shoulder (24.8 cm) and elbow (24.1 cm) were significantly greater than those for the hand (21.5 cm). Our analysis showed that speed and acclimation significantly affected the distance, although not gender or the side from which the approach came.

Based on the obtained results, we implemented a pre-touch reaction distance model in human–robot pre-touch interaction. We used an android robot with a human-like masculine appearance whose reaction was reflected in the defined reaction distances obtained in our human–human interaction analysis. Knowledge about upper body pre-touch reaction distance, based on human–human touch interaction and the differences of its effectiveness, will contribute to gathering pre-touch proxemics information for social robots that physically interact with people.

ACKNOWLEDGMENT

This work was supported by JST CREST Grant Number JPMJCR18A1, Japan.

REFERENCES

[1] K. M. Grewen, B. J. Anderson, S. S. Girdler, and K. C. Light, "Warm partner contact is related to lower cardiovascular reactivity," *Behavioral Medicine*, vol. 29, no. 3, pp. 123–130, 2003.

[2] S. Cohen, D. Janicki-Deverts, R. B. Turner, and W. J. Doyle, "Does hugging provide stress-buffering social support? A study of susceptibility to upper respiratory infection and illness," *Psychological Science*, vol. 26, no. 2, pp. 135–147, 2015.

[3] B. K. Jakubiak and B. C. Feeney, "Keep in touch: The effects of imagined touch support on stress and exploration," *Journal of Experimental Social Psychology*, vol. 65, pp. 59–67, 2016.

[4] A. Gallace and C. Spence, "The science of interpersonal touch: An overview," *Neuroscience & Biobehavioral Reviews*, vol. 34, no. 2, pp. 246–259, 2010.

[5] K. C. Light, K. M. Grewen, and J. A. Amico, "More frequent partner hugs and higher oxytocin levels are linked to lower blood pressure and heart rate in premenopausal women," *Biological Psychology*, vol. 69, no. 1, pp. 5–21, 2005.

[6] T. Field, "Touch for socioemotional and physical well-being: A review," *Developmental Review*, vol. 30, no. 4, pp. 367–383, 2010.

[7] R. Yu, E. Hui, J. Lee, D. Poon, A. Ng, K. Sit, K. Ip, F. Yeung, M. Wong, and T. Shibata, "Use of a therapeutic, socially assistive pet robot (PARO) in improving mood and stimulating social interaction and communication for people with dementia: Study protocol for a randomized controlled trial," *JMIR Research Protocols*, vol. 4, no. 2, p. e45, 2015.

[8] H. Sumioka, A. Nakae, R. Kanai, and H. Ishiguro, "Huggable communication medium decreases cortisol levels," *Scientific Reports*, vol. 3, p. 3034, 2013.

[9] M. Shiomi, K. Nakagawa, K. Shinozawa, R. Matsumura, H. Ishiguro, and N. Hagita, "Does a robot's touch encourage human effort?," *International Journal of Social Robotics*, vol. 9, pp. 5–15, 2016.

[10] M. Shiomi, A. Nakata, M. Kanbara, and N. Hagita, "A hug from a robot encourages prosocial behavior," In *2017 26th IEEE International Symposium on Robot and Hu-man Interactive Communication (RO-MAN)*, Lisbon, Portugal, pp. to appear, 2017.

[11] C. Bevan, and D. S. Fraser, "Shaking hands and cooperation in tele-present human-robot negotiation," In *Proceedings of the Tenth Annual ACM/IEEE International Conference on Human-Robot Interaction*, pp. 247–254, 2015.

[12] M. Shiomi, A. Nakata, M. Kanbara, and N. Hagita, "A robot that encourages self-disclosure by hug," In *Social Robotics: 9th International Conference, ICSR 2017*, Tsukuba, Japan, November 22–24, 2017, Proceedings, A. Kheddar, E. Yoshida, S. S. Ge et al., eds., pp. 324–333, Cham: Springer International Publishing, 2017.

[13] M. Shiomi and N. Hagita, "Do audio-visual stimuli change hug impressions?" In *Social Robotics: 9th International Conference, ICSR 2017*, Tsukuba, Japan, November 22–24, 2017, Proceedings, A. Kheddar, E. Yoshida, S. S. Ge et al., eds., pp. 345–354, Cham: Springer International Publishing, 2017.

[14] T. L. Chen, C.-H. A. King, A. L. Thomaz, and C. C. Kemp, "An investigation of responses to robot-initiated touch in a nursing context," *International Journal of Social Robotics*, vol. 6, no. 1, pp. 141–161, 2013.

[15] H. Fukuda, M. Shiomi, K. Nakagawa, and K. Ueda, "'Midas touch' in human-robot interaction: Evidence from event-related potentials during the ultimatum game," In *Human-Robot Interaction (HRI), 2012 7th ACM/IEEE International Conference on*, pp. 131–132, 2012.

[16] M. Shiomi, K. Shatani, T. Minato, and H. Ishiguro, "How should a robot react before people's touch?: Modeling a pre-touch reaction distance for a robot's face," *IEEE Robotics and Automation Letters*, vol. 3, no. 4, pp. 3773–3780, 2018.

[17] A. Saito, M. Kimoto, T. Iio, K. Shimohara, and M. Shiomi, "Preliminary investigation of pre-touch reaction distances toward virtual agents," In *Proceedings of the 7th International Conference on Human-Agent Interaction*, pp. 292–293, 2019.

[18] J. T. Suvilehto, E. Glerean, R. I. M. Dunbar, R. Hari, and L. Nummenmaa, "Topography of social touching depends on emotional bonds between humans," *Proceedings of the National Academy of Sciences*, vol. 112, no. 45, pp. 13811–13816, 2015.

[19] E. T. Hall, *The Hidden Dimension. Knopf Doubleday Publishing Group*. 1966.

[20] R. Kirby, R. Simmons, and J. Forlizzi, "Companion: A constraint-optimizing method for person-acceptable navigation," *In 2009 The 18th IEEE International Symposium on Robot and Human Interactive Communication, 2009. RO-MAN*, pp. 607–612, 2009.

[21] M. Luber, L. Spinello, J. Silva, and K. O. Arras, "Socially-aware robot navigation: A learning approach," In *2012 IEEE/RSJ International Conference on Intelligent Robots and Systems (IROS)*, pp. 902–907, 2012.

[22] M. Svenstrup, T. Bak, and H. J. Andersen, "Trajectory planning for robots in dynamic human environments," In *2010 IEEE/RSJ International Conference on Intelligent Robots and Systems (IROS)*, pp. 4293–4298, 2010.

[23] J. V. Gómez, N. Mavridis, and S. Garrido, "Social path planning: Generic human-robot interaction framework for robotic navigation tasks," In *2nd International Workshop on Cognitive Robotics Systems: Replicating Human Actions and Activities*, pp. 17–18, 2013.

[24] S. Satake, T. Kanda, D. F. Glas, M. Imai, H. Ishiguro, and N. Hagita, "A robot that approaches pedestrians," *IEEE Transactions on Robotics*, vol. 29, no. 2, pp. 508–524, 2013.

[25] C.-M. Huang, T. Iio, S. Satake, and T. Kanda, "Modeling and controlling friendliness for an interactive museum robot," In *Robotics: Science and Systems*, pp. 12–16, 2014.

[26] L. Takayama and C. Pantofaru, "Influences on proxemic behaviors in human-robot interaction," In *2009 IEEE/RSJ International Conference on Intelligent Robots and Systems*, pp. 5495–5502, 2009.

[27] J. Mumm and B. Mutlu, "Human-robot proxemics: Physical and psychological distancing in human-robot interaction," In *Proceedings of the 6th International Conference on Human-Robot Interaction*, pp. 331–338, 2011.

[28] S. Rossi, M. Staffa, L. Bove, R. Capasso, and G. Ercolano, "User's personality and activity influence on HRI comfortable distances," In *International Conference on Social Robotics*, pp. 167–177, 2017.

[29] M. Shiomi, H. Sumioka, K. Sakai, T. Funayama, and T. Minato, "SŌTO: An android platform with a masculine appearance for social touch interaction," In *Companion of the 2020 ACM/IEEE International Conference on Human-Robot Interaction,* Cambridge, United Kingdom, pp. 447–449, 2020.

[30] D. A. Cuello Mejía, H. Sumioka, H. Ishiguro, and M. Shiomi, "Modeling a pre-touch reaction distance around socially touchable upper body parts of a robot," *Applied Sciences,* vol. 11, no. 16, p. 7307, 2021.

[31] M. Baldassare and S. Feller, "Cultural variations in personal space," *Ethos,* vol. 3, no. 4, pp. 481–503, 1975.

[32] S. Heshka and Y. Nelson, "Interpersonal speaking distance as a function of age, sex, and relationship," *Sociometry,* vol. 35, pp. 491–498, 1972.

[33] M. L. Knapp, J. A. Hall, and T. G. Horgan, *Nonverbal Communication in Human Interaction.* Cengage Learning: Wadsworth Pub Co., 2013.

[34] E. McDaniel and P. A. Andersen, "International patterns of interpersonal tactile communication: A field study," *Journal of Nonverbal Behavior,* vol. 22, no. 1, pp. 59–75, 1998.

[35] R. Dibiase and J. Gunnoe, "Gender and culture differences in touching behavior," *The Journal of Social Psychology,* vol. 144, no. 1, pp. 49–62, 2004.

[36] J. A. Hall, "Touch, status, and gender at professional meetings," *Journal of Nonverbal Behavior,* vol. 20, no. 1, pp. 23–44, 1996.

[37] F. N. Willis and L. F. Briggs, "Relationship and touch in public settings," *Journal of Nonverbal Behavior,* vol. 16, no. 1, pp. 55–63, 1992.

[38] D. F. Glas, T. Minato, C. T. Ishi, T. Kawahara, and H. Ishiguro, "Erica: The erato intelligent conversational android," In *2016 25th IEEE International Symposium on Robot and Human Interactive Communication (RO-MAN),* pp. 22–29, 2016.

[39] T. Kanda, H. Ishiguro, T. Ono, M. Imai, and R. Nakatsu, "Development and evaluation of an interactive humanoid robot "Robovie"," In *Proceedings 2002 IEEE International Conference on Robotics and Automation (Cat. No.02CH37292),* vol. 2, pp. 1848–1855, 2002.

[40] R. Matsumura, M. Shiomi, K. Nakagawa, K. Shinozawa, and T. Miyashita, "A desktop-sized communication robot: "Robovie-mr2"," *Journal of Robotics and Mechatronics,* vol. 28, no. 1, pp. 107–108, 2016.

[41] R. Matsumura and M. Shiomi, "An animation character robot that increases sales," *Applied Sciences,* vol. 12, no. 3, p. 1724, 2022.

[42] H. Sumioka, N. Yamato, M. Shiomi, and H. Ishiguro, "A minimal design of a human infant presence: A case study toward interactive doll therapy for older adults with dementia," *Frontiers in Robotics and AI,* vol. 8, p. 164, 2021.

[43] V. Mehta, "The new proxemics: COVID-19, social distancing, and sociable space," *Journal of Urban Design,* vol. 25, no. 6, pp. 669–674, 2020.

[44] R. Welsch, M. Wessels, C. Bernhard, S. Thönes, and C. von Castell, "Physical distancing and the perception of interpersonal distance in the COVID-19 crisis," *Scientific Reports,* vol. 11, no. 1, pp. 1–9, 2021.

[45] M. Shiomi, A. Kubota, M. Kimoto, T. Iio, and K. Shimohara, "Stay away from me: Coughing increases social distance even in a virtual environment," *PLoS One,* vol. 17, no. 12, p. e0279717, 2022.

CHAPTER 8

Comparison of Pre-touch Reaction Distance between Physical and VR Environments

Dario Alfonso Cuello Mejía, Aoba Saito,

Mitsuhiko Kimoto, Takamasa Iio,

Katsunori Shimohara, Hidenobu Sumioka,

Hiroshi Ishiguro, and Masahiro Shiomi

Advanced Telecommunications Research
Institute International, Kyoto, Japan

Osaka University, Osaka, Japan

Doshisha University, Kyoto, Japan

8.1 INTRODUCTION

Research on the nature of interactions in virtual reality (VR) environments has progressed in recent years thanks to the development of new devices that introduce tactile stimuli for a more natural flow of multisensory information [1] and algorithms that effectively process tactile stimuli using the human cognition of the sense of touch [2]. Advances in this field are fuelling potential interaction with agents in VR environments using real physical stimuli [3–7].

DOI: 10.1201/9781003384274-11

However, prior research has focused mainly on the touch and after-touch interactions of virtual agents with participants without incorporating any of the factors involved in the before-touch interactions. One study concluded that if a robot's reaction distance before being touched were to resemble that of humans, this behaviour would help convey a more natural and human-like impression [8]. In a VR context, simulating interactions between objects and agents based on physical phenomenon has been studied [1], although again the reaction distance before being touched was ignored. Although some research has focused on human-agent proxemics, e.g., personal space in VR [9–11] and human proxemics preferences with regard to robots [12], none of these studies examined pre-touch situations.

We address this issue by measuring the pre-touch distance in a VR environment, which is defined as the distance at which a person usually reacts before being touched. First, we collected data on the distance at which participants began to feel uncomfortable when a virtual agent tries to touch their face. Then, we analyzed the obtained data and defined the characteristics of the optimal pre-touch distance for VR spaces. Next, we tested how the interactions can be improved by implementing pre-touch distance behaviour when a participant tries to touch the agent in a VR environment (Figure 8.1). Note that this chapter is modified based on our previous work [13], edited to be comprehensive and fit with the context of this book.

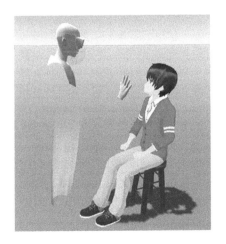 

FIGURE 8.1 Pre-touch interaction with an agent.

8.2 MEASUREMENT OF PRE-TOUCH DISTANCE TOWARDS VIRTUAL AGENTS

8.2.1 Data Collection

For the data collection, we measured the distance at which a participant began to feel uncomfortable when a virtual agent extended its hand towards his/her face. Our method of measuring this distance is based on the stop distance paradigm [14–19]. In this method, a virtual agent gradually moves its hand closer to the participant's face, and the distance at which the participant starts to feel uncomfortable is recorded. The pre-touch distance in VR space is calculated as the distance between the end of the agent's right-hand middle finger and the participant's viewpoint.

8.2.2 Experimental Setup

We used Unity, which is a game development platform, to deploy a 3D model of the agents, their movements, and the virtual environment. We also implemented functions for stopping the approach of the agent's hand and measuring the distance between the participant and the agent. We used Oculus Rift, an HMD for virtual reality, for the implementation. By linking it to Unity, the HMD's position and orientation are reflected in a VR space, and people can feel as if they are synchronously in a VR space. We used an Xbox controller as the user interface to stop the agent's hand progress towards the participant's face. Participants could press the button at any time, an action that immediately recorded the pre-touch distance.

We investigated the effect of the avatar's gender on the participants' pre-touch behaviour using two animated 3D models (from the Unity Assets Store) as agents: a man and a woman. The touching behaviour was defined as follows: the participant is sitting, and the agent is standing and extends her/his right hand towards the participant's face from various angles (Figure 8.2). Considering the direction of the participant's face, there are 19 vertical angles and 13 horizontal ones. We set 247 angles in random order.

The agent's hand, which always started 70 cm from the participant's face at every angle, was programmed to approach it as quickly as possible within five seconds, which is a constant movement speed of 14 cm/s. We designed the touching motion for the agent's palm to be the closest to the participant's face by adjusting the approaching angle of the touching motion. The initialization of the touching behaviour and the agents' hand position was set with a button on the participant's controller.

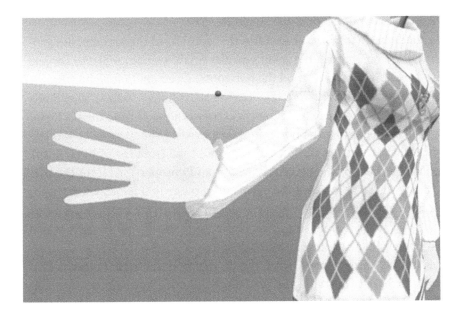

FIGURE 8.2　Hand's approach.

8.2.3　Experiment Procedure

Twenty individuals participated in the data collection (ten men and ten women, aged 18–24). They wore the HMD and listened to our explanation of the experiment. When they pressed the controller's button, an agent appeared whose right hand's position was fixed. After confirming the hand's position, the participants pressed the again button to initiate the agent's touching motion. They stopped the agent's hand by pressing the same button when they began to feel uncomfortable. Then, the system measured the distance between the participant's face and the palm of the agent's hand. After finishing the measurement procedure for all the target angles, participants were given a three-minute break, and then the agent's gender was changed and the measurements were repeated.

8.2.4　Data Analysis

In the acquired dataset, we measured 4,934 data points for the men (with male agents: 2,467 points; with female agents: 2,467 points) and 4,918 points for the women (with male agents: 2,457 points; with female agents: 2,461 points). The average pre-touch distance for all the participants was 17.93 cm ($SE = 1.98$ cm). We analyzed the data with a focus on gender, angles, and habituation effects.

8.2.4.1 Gender and Approach Angle Effects

We conducted a four-factor ANOVA that considered the participant's gender (man/woman, PG), the agent's gender (man/woman, AG), the vertical direction (high/low, HL), and the horizontal direction (left/right, LR).

We found significant differences in the HL factor ($p=0.024$), the LR factor ($p=0.019$), and the interaction effect between the AG and the HL factor ($p=0.010$). Regarding the horizontal factors, the pre-touch space was significantly longer from the left side (18.4 cm) than from the right (17.4 cm), perhaps because of the perceived agent's body proportions from the participant's visual field. Since the agents used their right hands when touching from the left side, their bodies were in front of the participant; when touching from the right side, their bodies were mostly out of sight, which may have influenced the pre-touch distance.

Multiple comparisons with the Bonferroni method revealed that when the agent was man, the pre-touch distance was significantly longer when the agent approached from below than from above: $p=0.010$, 19.3 cm for the lower side and 16.8 cm for the upper side. These results indicate that the touch angle, in combination with the agent's gender, influenced the pre-touch distance.

8.2.4.2 Habituation Effect

We conducted a three-factor analysis considering the participant's gender (men/women), the agent's gender (men/women), and data collection duration (first ten/final ten pre-touch interaction). Significant differences were found in the interactions of the participant's gender, the agent's gender, and the time factors ($p=0.043$). Multiple comparisons with the Bonferroni method revealed that the first ten trials showed a significantly longer distance than the final ten trials when the participant was man and the agent was woman: $p=0.028$, 22.3 cm for the first ten trials and 18 cm for the final ten.

8.2.5 Pre-touch Distance Trending

For evaluating how to classify the pre-touch distance based on the average distance (17.93 cm), we performed clustering by the k-means method with two clusters: near (less than average) and far (above average). Results showed that the average pre-touch distance obtained for the near group was 9.24 and 25.03 cm for the far group, suggesting that the pre-touch distance preferred by humans is not necessarily uniform. Therefore, in the second phase of our study, we used the pre-touch distance mean of each group to determine the proper reaction of an agent to a touch behaviour.

8.3 EVALUATION OF AGENT'S RESPONSE TOWARDS TOUCH ATTEMPTS

8.3.1 Experimental Setup and Conditions

For this experiment, we considered the pre-touch distance (Near/Average/Far), the agent's gender factor (man/woman), and the participant's gender factor (man/woman). The following are the details of each condition in the pre-touch distance factor:

- Near condition: The near group average obtained in the data collection (9.24 cm).

- Average condition: The average of all the participants in the data collection (17.93 cm).

- Far condition: The far group average obtained in the data collection (25.03 cm).

For the experimental environment, we created a VR environment where the participants performed a touch attempt towards the agent. We attached a wireless controller to their right hand to reflect their position in the physical space in the VR space, and the participants extended their hands towards the virtual agent's face. Both the participants and their avatars were standing, while the virtual agent was sitting.

8.3.2 Procedure

Twenty-eight participants (14 men and 14 women aged 18–24) joined this experiment. First, they filled out a consent form. Then, they put on and adjusted the HMD while listening to an explanation of the procedure. A simplified version of the data collection procedure was performed to obtain the pre-touch distance for each participant. After collecting such data, we started the experiment. A man or woman agent sat in the VR space. The participant could move his/her right hand towards the agent's face at any time. The agent looked at the participant's face when his/her hand approached within a certain distance from the agent's face. We measured this distance between the participant's hand and the agent's face. We explained to the participants that this action indicated restlessness or discomfort and asked them to stop their approach immediately when the agent's reaction was observed.

Under each condition, the distance at which the agent responded to the participant's hand was determined based on the three conditions defined above. Participants moved their hands towards the agent's face nine times from different angles and observed their responses. After the experiment, they filled out a questionnaire about their preferences. This procedure was performed for all the pre-touch distance conditions, and after a short break, it was repeated with an agent of a different gender.

8.3.3 Measurement and Results

We examined the following three topics in questionnaires and interviews to gauge our participants' perceptions:

- Participants' likeability towards the agent: Using the likeability item of the Godspeed questionnaire series [20], participants evaluated the agents' friendliness on a 1-to-7 scale with 1 being mostly negative.

- Perceived agent's likeability towards participants: We prepared items to estimate the perceived agent's feelings of likeability, closeness, and friendliness towards the participants based on the likeability items of Godspeed: like-dislike, friendly-unfriendly, kind-unkind, pleasant-unpleasant, and nice-awful. Each item was estimated with 1-to-7 scale questions with 1 being mostly negative.

- Pre-touch distance match: After the experiment was completed, we interviewed the participants to check which of the reaction distance conditions was preferred, ranking them from the most to least liked. We compared the interview results with the pre-touch distance group (Near/Far) of the participants obtained during data collection.

8.3.4 Results

Our analysis of the data collection before the experiment placed 14 participants in the near group (seven men and seven women) and 14 participants in the far group (seven men and seven women). The classification was based on the average values from the data collection before the experiment for each participant.

8.3.4.1 Likeability towards the Agent

We performed a four-factor mixed ANOVA (agent pre-touch distance (AD), participant pre-touch distance group (PD), agent gender (AG), and participant gender (PG)) on the participants' likeability questionnaire results (Table 8.1). We found a significant difference in the pre-touch distance factor of the agents. The results obtained with multiple comparisons

TABLE 8.1   Pre-touch Distance and Preference of Participants
(*: $p<0.05$, +: $p<0.10$)

		Pre-touch Distance Groups (Participants)	
		Near	Far
	Near	5*	0*
Pre-touch distance conditions (agent)	Average	7	8
	Far	2+	6+

on the agents' pre-touch distance showed a significant trend between the near and far conditions ($p=0.083$), and that the average condition was significantly higher than the far condition (Average > Far: $p=0.012$). There was no significant difference between the near and average conditions ($p=1.000$).

8.3.4.2 Participant's Perceived Likeability

We also performed a four-factor mixed ANOVA (agent pre-touch distance (AD), participant pre-touch distance group (PD), agent gender (AG), and participant gender (PG)) on the agents' likeability questionnaire results. Similar to the participant's likeability, we found a significant difference in the agent's pre-touch distance factor ($p=0.014$). The results of multiple comparisons on the agent's pre-touch distance showed that the near condition had significantly higher values than the far condition (Near > Far: $p<0.001$), the near condition had significantly higher values than the average condition (Near > Average: $p=0.002$), and the average condition had significantly higher values than the far condition (Average > Far: $p<0.001$).

8.3.4.3 Participant's Reaction Distance Preferences

Table 8.1 shows the ratio of the participants' pre-touch distance group (Near/Far) and the preferred agents' pre-touch reaction distance (Near/Average/Far). A chi-square test showed a significant difference among the conditions ($x^2(2)=7.067$, $p=0.029$). A residual analysis showed that the near distance group significantly preferred the near agent's reaction distance over the far one. We also found a significant trend in the far distance group, which preferred the far agent's reaction distance over the near one.

8.4 DISCUSSION

8.4.1 Application of Pre-touch Interaction Findings in VR Space into Physical Space

The results of the data collection in the VR space in this study resemble those in the physical space, although differences did emerge. If we can approximate the pre-touch distance for the physical space by adding certain coefficients to the pre-touch distance in the VR space, we can probably approximate a pre-touch distance to a body part that would be difficult to measure in reality (including from an ethical point of view) by implementing the conditions in a VR environment. In addition, in VR space, it is easier to set and modify the appearance, arm trajectory, speed, and other characteristics of an agent that is trying to touch a human than in a physical space. Therefore, this research contributed to the development of a pre-touch distance model that takes into account various factors of the human body.

A previous study on pre-touch interaction in a physical space calculated the overall mean of the pre-touch distance without distinguishing between near/far groups [8]. Therefore, when we implemented our analysis from a current study using the previous study's data, the generated clusters and averages were different for the below and above average groups (Figure 8.3).

8.4.2 Distribution of Pre-touch Distance in Physical and VR Spaces

When we compared the data from both studies, the pre-touch distance in the physical space was slightly longer than that in VR space. For the average and far groups, distances in the physical space were longer than those in the VR space by about 2 cm. In the near group, the distance difference was about 5 cm. On the other hand, we also identified common aspects between VR and physical spaces. The preferred pre-touch distance in both lengthened as we got closer to the face area. In other words, the greater the distance, the closer was the pre-touch distance between VR and physical spaces.

To investigate the different distribution of pre-touch distance between different spaces, using the same appearance robots/agents will be useful. In the context of personal distance, a past study used Pepper robot/agent to investigate the difference in the personal distance between physical and VR spaces [12]. Similar to the past study, using virtual agents with different appearances of robots (e.g., human-like appearances [21,22] and mechanical-robot-like appearances [23–25]) would be an interesting future work.

FIGURE 8.3   Clustering pre-touch distance based on mean.

8.5 CONCLUSION

In this study, we first collected data on pre-touch behaviour to investigate the potential interaction in a VR space and found that the pre-touch distance changes depending on the agent's appearance and its touch angle. We also found that the pre-touch distance measured in physical space resembled that obtained in VR space. Similarly, the pre-touch distance in both spaces can be classified into two types: near and far.

Next, we experimentally determined the participants' perception in a touch interaction with an agent in the VR space. The results showed that an agent that responds at an average pre-touch distance conveys a more familiar and friendlier impression than an agent that reacts at a longer pre-touch distance; the nearer the agent's pre-touch distance is, the friendlier it appears. We also clarified that a person with a nearer pre-touch distance prefers an agent with a similar close pre-touch distance. The same is true for a person with a longer pre-touch distance.

ACKNOWLEDGEMENT

This work was supported by JST CREST Grant Number JPMJCR18A1, Japan.

REFERENCES

[1] G. Huisman, "Social touch technology: A survey of haptic technology for social touch," *IEEE Transactions on Haptics*, vol. 10, no. 3, pp. 391–408, 2017.

[2] J. B. F. van Erp, and A. Toet, "Social touch in human-computer interaction," *Frontiers in Digital Humanities*, vol. 2, no. 2, p. 2, 2015.

[3] J. N. Bailenson, and N. Yee, "Virtual interpersonal touch: Haptic interaction and copresence in collaborative virtual environments," *Multimedia Tools and Applications*, vol. 37, no. 1, pp. 5–14, 2008.

[4] G. Huisman, M. Bruijnes, J. Kolkmeier, M. Jung, A. D. Frederiks, and Y. Rybarczyk, "Touching virtual agents: Embodiment and mind," In *International Summer Workshop on Multimodal Interfaces*, pp. 114–138, 2013.

[5] M. Shiomi, and N. Hagita, "Audio-visual stimuli change not only robot's hug impressions but also its stress-buffering effects," *International Journal of Social Robotics*, vol. 13, pp. 469–476, 2019.

[6] K. Higashino, M. Kimoto, T. Iio, K. Shimohara, and M. Shiomi, "Tactile stimulus is essential to increase motivation for touch interaction in virtual environment," *Advanced Robotics*, vol. 35, no. 17, pp. 1043–1053, 2021.

[7] M. Kimoto, Y. Otsuka, M. Imai, and M. Shiomi, "Effects of appearance and gender on pre-touch proxemics in virtual reality," *Frontiers in Psychology*, vol. 14, p. 1195059, 2023.

[8] M. Shiomi, K. Shatani, T. Minato, and H. Ishiguro, "How should a robot react before people's touch?: Modeling a pre-touch reaction distance for a robot's face," *IEEE Robotics and Automation Letters*, vol. 3, no. 4, pp. 3773–3780, 2018.

[9] F. A. Sanz, A.-H. Olivier, G. Bruder, J. Pettré, and A. Lécuyer, "Virtual proxemics: Locomotion in the presence of obstacles in large immersive projection environments," In *2015 IEEE Virtual Reality (VR)*, pp. 75–80, 2015.

[10] M. Lee, G. Bruder, T. Höllerer, and G. Welch, "Effects of unaugmented periphery and vibrotactile feedback on proxemics with virtual humans in AR," *IEEE Transactions on Visualization Computer Graphics*, vol. 24, no. 4, pp. 1525–1534, 2018.

[11] M. Shiomi, A. Kubota, M. Kimoto, T. Iio, and K. Shimohara, "Stay away from me: Coughing increases social distance even in a virtual environment," *PLoS One*, vol. 17, no. 12, p. e0279717, 2022.

[12] R. Li, M. van Almkerk, S. van Waveren, E. Carter, and I. Leite, "Comparing human-robot proxemics between virtual reality and the real world," In *2019 14th ACM/IEEE International Conference on Human-Robot Interaction (HRI)*, pp. 431–439, 2019.

[13] D. A. C. Mejía, A. Saito, M. Kimoto, T. Iio, K. Shimohara, H. Sumioka, H. Ishiguro, and M. Shiomi, "Modeling of pre-touch reaction distance for faces in a virtual environment," *Journal of Information Processing*, vol. 29, pp. 657–666, 2021.

[14] L. Takayama, and C. Pantofaru, "Influences on proxemic behaviors in human-robot interaction," In *2009 IEEE/RSJ International Conference on Intelligent Robots and Systems,* pp. 5495–5502, 2009.

[15] J. Mumm, and B. Mutlu, "Human-robot proxemics: physical and psychological distancing in human-robot interaction," In *Proceedings of the 6th International Conference on Human-Robot Interaction,* pp. 331–338, 2011.

[16] M. Obaid, E. B. Sandoval, J. Złotowski, E. Moltchanova, C. A. Basedow, and C. Bartneck, "Stop! That is close enough. How body postures influence human-robot proximity," In *2016 25th IEEE International Symposium on Robot and Human Interactive Communication (RO-MAN),* pp. 354–361, 2016.

[17] S. Rossi, M. Staffa, L. Bove, R. Capasso, and G. Ercolano, "User's personality and activity influence on HRI comfortable distances," In *International Conference on Social Robotics,* pp. 167–177, 2017.

[18] Y. Hiroi, and A. Ito, "Influence of the height of a robot on comfortableness of verbal interaction," *IAENG International Journal of Computer Science,* vol. 43, no. 4, pp. 447–455, 2016.

[19] Y. Kim, S. S. Kwak, and M.-S. Kim, "Am I acceptable to you? Effect of a robot's verbal language forms on people's social distance from robots," *Computers in Human Behavior,* vol. 29, no. 3, pp. 1091–1101, 2013.

[20] C. Bartneck, D. Kulić, E. Croft, and S. Zoghbi, "Measurement instruments for the anthropomorphism, animacy, likeability, perceived intelligence, and perceived safety of robots," *International Journal of Social Robotics,* vol. 1, no. 1, pp. 71–81, 2009.

[21] D. F. Glas, T. Minato, C. T. Ishi, T. Kawahara, and H. Ishiguro, "Erica: The erato intelligent conversational android," In *2016 25th IEEE International Symposium on Robot and Human Interactive Communication (RO-MAN),* pp. 22–29, 2016.

[22] M. Shiomi, H. Sumioka, K. Sakai, T. Funayama, and T. Minato, "SŌTO: An android platform with a masculine appearance for social touch interaction," In *Companion of the 2020 ACM/IEEE International Conference on Human-Robot Interaction, Cambridge, United Kingdom,* pp. 447–449, 2020.

[23] T. Kanda, H. Ishiguro, T. Ono, M. Imai, and R. Nakatsu, "Development and evaluation of an interactive humanoid robot "Robovie"," In *Proceedings 2002 IEEE International Conference on Robotics and Automation (Cat. No.02CH37292),* vol. 2, pp. 1848–1855, 2002.

[24] R. Matsumura, M. Shiomi, K. Nakagawa, K. Shinozawa, and T. Miyashita, "A desktop-sized communication robot: "Robovie-mr2"," *Journal of Robotics and Mechatronics,* vol. 28, no. 1, pp. 107–108, 2016.

[25] R. Matsumura, and M. Shiomi, "An animation character robot that increases sales," *Applied Sciences,* vol. 12, no. 3, p. 1724, 2022.

SECTION 4

**Interaction Design for Touching
and Being Touched**

CHAPTER 9

Understanding Natural Reaction Time toward Touch

Atsumu Kubota, Mitsuhiko Kimoto,
Takamasa Iio, Katsunori Shimohara,
and Masahiro Shiomi

*Advanced Telecommunications Research
Institute International, Kyoto, Japan*

Doshisha University, Kyoto, Japan

9.1 INTRODUCTION

People usually prefer a quick response time (less than one second) when they use computers [1,2]. A past study recommended a similar principle as the two-second rule, i.e., computer systems should be designed to respond within two seconds to avoid hesitation from users [3]. In fact, recent computer systems such as web servers are designed to respond to user requests as quickly as possible.

However, an overly quick response time is inappropriate in human-like interaction settings such as conversation. For example, in human–human conversations, people often make a short pause during their speech, especially in turn-taking situations. In other words, addressing human-like reaction time is critical for achieving natural and smooth interactions for social robots when they interact with humans.

DOI: 10.1201/9781003384274-13

103

Therefore, robotics researchers have broadly investigated natural reaction time design in the context of human–robot interaction: listening to people's speech [4], conveying emotions [5,6], reaction behaviors [7,8], and conversations [9,10]. Most of these studies generally concluded that an appropriate reaction time is less than one second, although such a "hasty" response time as practically zero seconds is not better due to different impressions from people's usual reaction times. Such basic knowledge has already been applied to social robots that work in real environments to provide various conversational services to customers [11–13].

However, although such timing behaviors are essential for natural interaction with people, detailed reaction time design is less focused on touch interaction. A few studies investigated the appropriate reaction time when robots are touched by people, although just with a human-like appearance robot [7,8]. Moreover, the resolutions of the reaction time comparison are relatively large, and neither study separated the reaction speeds or the start timings. Both studies also claimed that a one-second reaction is superior to zero seconds in a conversational setting, and the latter reported that people preferred non-zero second reaction times without providing any appropriate detailed timing. Therefore, the precise reaction time design toward touch considering different types of robots remains unknown. Based on these considerations, we investigated individually appropriate reaction times when different types of robots are touched and with shorter time resolutions. Since many comparisons are needed, we conducted a web-based survey experiment to measure the variety of reaction behaviors with different timings and speeds.

One unique point of this study is that we investigated appropriate reaction times by addressing two different reaction times by separating the reaction behaviors into two stages: (1) the time length between being touched and the start of a reaction behavior (before-reaction time) and (2) the time lengths of the reaction behavior (after-reaction time) (Figure 9.1). Previous studies did not separate the reaction behaviors toward touching, even though this process enables social robots to achieve more natural reaction behaviors. For example, if robotics researchers were to design a reaction behavior within one second, times of 800-ms before-reaction and 200-ms after-reaction create different feelings compared to times of 200-ms before-reaction and 800-ms after-reaction. Note that this chapter is modified based on our previous work [14], edited to be comprehensive and fit with the context of this book.

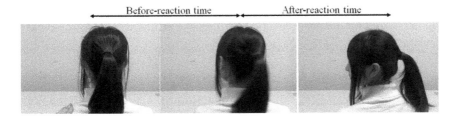

FIGURE 9.1   Before-reaction and after-reaction time concept.

9.2 MATERIALS AND METHODS OF EXPERIMENT

9.2.1 Models of Video Stimulus

To investigate the appropriate reaction time toward touching from various viewpoints, we prepared five models: three different types of robots (Sota/Nao/Pepper) and two different human models (man/woman). In all the videos, we placed the models on the right side of the videos, and the experimenter's hand touched the left shoulder of the models. The detailed behavior design is described in the next subsection. Here, we show a part of the video stimulus of each model.

Sota: Sota (VSTONE), which has eight degrees of freedom (DOFs): three DOFs for its head, two DOFs for both arms, and one DOF for its lower body. It is 28 cm tall.

Nao: Nao (Aldebaran Robotics), which has 25 DOFs: two DOFs for its head, six DOFs for both arms, and 11 DOFs for its lower body. It is 58 cm tall.

Pepper: Pepper (Softbank Robotics), which has 20 DOFs: two DOFs for its head, six DOFs for both arms, and six DOFs for its lower body. It is 121 cm tall.

Humans: We recruited both a man and a woman whose heights are about 172 and 153 cm.

9.2.2 Basic Settings of Visual Stimulus

For the preparation of these video stimuli, we set the length of each video to five seconds without any sound. The widths and heights of each video were 1,280 and 720 pixels, respectively, and the fps values of each were 29.97. Since we compressed the video files to decrease their total size (all the videos are less than 1 MB), the participants could watch them after completely downloading them. An operator controlled the robot's reaction behaviors for accurate reaction timing. We edited the videos to avoid acceleration effects of the reaction behaviors.

9.2.3 Before-Reaction Time

To investigate the effects of the before-reaction time, we used 200 ms as a time-slice resolution in the video stimulus because past studies reported that reaction times to touch stimuli range between 150 and 400 ms [15–19]. We used 200 ms as a relatively large resolution because using a very short time resolution (e.g., 33 ms) would create too many conditions and cause difficulties when applying various kinds of robots, especially a cheaper one that has relatively low-spec servo motors and processing capabilities.

The range of searching before-reaction times is between zero and one second in 200-ms intervals (i.e., 0, 200, 400, 600, 800, and 1,000 ms). We investigated the effect of zero seconds because we assume that a robot's sensing systems enable them to estimate the actual contact timing, and such prediction-based reaction behaviors may change people's impressions.

9.2.4 After-Reaction Time

We also used 200 ms as the time-slice resolution in the video stimulus to investigate the effects of the after-reaction times. One different setting is to ignore the effects of zero seconds because finishing a reaction behavior within zero seconds is obviously impossible. Thus, the ranges of searching the after-reaction times are between 200 ms and one second in 200-ms intervals (i.e., 200, 400, 600, 800, and 1,000 ms).

9.2.5 Procedure

Due to the need for large numbers of comparisons, we conducted a web-based survey with participants recruited by a Japanese survey company. They first accessed a web page that displayed explanatory texts of the data collection and how to evaluate each video. Those who agreed to the terms of our experimental procedure gained access to the survey pages. Before the video stimuli, we explained that the experiment was investigating natural reaction times in response to different touch behaviors.

In the web survey, we prepared 30 videos for each model (six before-reaction times and five after-reaction times). We also prepared two additional videos (zero seconds in both before- and after-reaction times) for each model to investigate whether the participants carefully watched them and to check the quality of their answers. We prepared such dummy questions to screen participants who provided poor answers [20].

Therefore, the participants watched 32 videos for each model and evaluated the naturalness of the reaction time for each one.

We determined the order of the videos in advance due to the large number of video combinations. To avoid biases, we sorted the video orders to avoid continuing similar before- and after-reaction times. However, we fixed the order of the two videos with zero before- and after-reaction times to the 16th and 32nd to check the quality of their answers.

9.2.6 Measurements

We just measured one questionnaire item to investigate the perceived reaction behavior's naturalness on a one to ten scale, where one is very unnatural and ten is very natural. This answer is also used in the filtering process; we only used the data of participants who correctly rated the two videos with zero before- and after-reaction times.

9.2.7 Participants

Our survey had 1,212 participants: Sota: 266, Nao: 242, Pepper: 266, man: 219, woman: 219. Filtering lowered the valid data number to 780: 64.4%, Sota: 143, Nao: 137, Pepper: 159, man: 166, woman: 175.

9.3 RESULTS AND DISCUSSION

9.3.1 Best Combinations between Before-and After-Reaction Times

Figure 9.2 shows the experiment results. The bold part indicates the highest value in the tables. We analyzed the effects of the before-reaction times, the after-reaction times, and the model factor toward the perceived naturalness by a three-factor ANOVA. The results showed a significant effect for all factors. There were significant differences in the before-reaction time (F (5, 3,870) = 231.735, $p < 0.001$, partial $\eta^2 = 0.230$), in the after-reaction time (F (4, 3,096) = 1,112.153, $p < 0.001$, partial $\eta^2 = 0.590$), in the model (F (4, 774) = 5.120, $p < 0.001$, partial $\eta^2 = 0.026$), in the simple interaction effects between the before- and after-reaction times (F (20, 15,480) = 53.198, $p < 0.001$, partial $\eta^2 = 0.064$), in the simple interaction effects between the before-reaction and the model (F (20, 15,480) = 5.129, $p < 0.001$, *partial* $\eta^2 = 0.026$), in the simple interaction effects between the after-reaction and the model (F (16, 15,480) = 74.044, $p < 0.001$, partial $\eta^2 = 0.277$), and in the two-way interaction effect (F (80, 15,480) = 3.127, $p < 0.001$, partial $\eta^2 = 0.016$). Due to such a large number of combinations, we are unable to describe the details of the interaction effects.

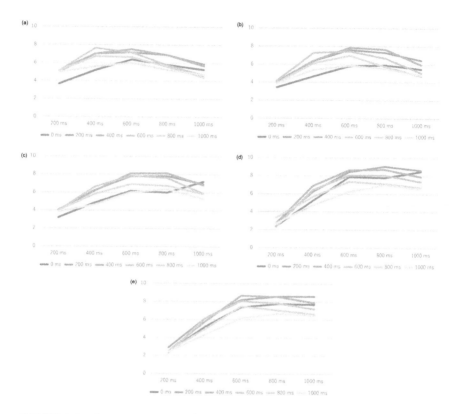

FIGURE 9.2 Questionnaire results of perceived naturalness. The vertical axis showed the questionnaire results, and the horizontal axis showed after-reaction time. Each line showed the different before-reaction time.

As shown in these results, each setting has only one peak, and the best combination between the before- and after-reaction times is different. The trends of the best combinations seem related to their anticipated body size. For example, slow before-reaction times and fast after-reaction times are preferred for Sota, and fast before-reaction times and slow after-reaction times are preferred for Pepper. In the human models, slow before-reaction times and fast after-reaction times are preferred for the woman model, whereas fast before-reaction times and slow after-reaction times are preferred for the man model.

9.3.2 Which Combinations Are Better for Robots?

As shown above, the experiment results found that the best combination of parameters is different among the models. Next, we investigated whether using the extracted parameters from observing human behaviors is the best approach for deciding the parameters of the robot behaviors. We compared the perceived naturalness when the robot uses the best combinations of human and robot models.

Figure 9.3 shows the results of comparisons of perceived naturalness with different parameter combinations. In the analysis of Sota's data, we conducted a repeated measure ANOVA, and its results showed a significant difference (F (3, 426)=7.437, $p<0.001$, partial $\eta^2=0.050$). Multiple comparisons with the Bonferroni method also showed significant differences: Sota's timing>man's timing ($p=0.003$) and>Pepper's timing ($p=0.02$). We found no significant differences between Sota's timing and that of Nao or the woman ($p=1.000$).

In the analysis of Nao's data, we conducted a repeated measure ANOVA and its results showed a significant difference (F (3, 408)=4.385, $p=0.005$, partial $\eta^2=0.031$). Multiple comparisons with the Bonferroni method showed significant differences: Nao's and the woman's timing>Sota's timing ($p=0.002$) and the man's timing ($p=0.037$). We found no significant differences with Pepper's timing ($p=1.000$).

In the analysis of Pepper's data, we conducted a repeated measure ANOVA, and its results showed a significant difference (F (3, 474)=27.935, $p<0.001$, partial $\eta^2=0.150$). Multiple comparisons with the Bonferroni method showed significant differences: Pepper's>Sota's timing ($p<0.001$). We found no significant differences in timing between Pepper and the man ($p=0.346$) or between Nao and the woman ($p=1.000$).

Our statistical analysis showed that using the individual best combination is optimal for increasing the perceived naturalness, which indicates the importance of parameter calibration. On the other hand, since investigating the best combinations might require too many tasks, investigating better combinations for different kinds of robots is also important. The experiment results showed that the best combinations of the Nao/woman model did not show significant disadvantages in any of the robots. Although an absence of evidence is not evidence of absence, using 400-ms before-reaction times and 600-ms after-reaction times would probably improve the natural reaction behavior of several robots.

As a reference, we conducted a non-inferiority analysis for the data of both Sota and Pepper to investigate the effectiveness of applying Nao's/woman's best combination for the reaction time. We set the non-inferiority margin at 10% in a one-sided test where $\alpha=0.025$. The analysis showed that the best combination for Nao and the woman (mean: 7.65, confidence interval (CI): 7.28 to 8.29) was non-inferior to Sota's best combination

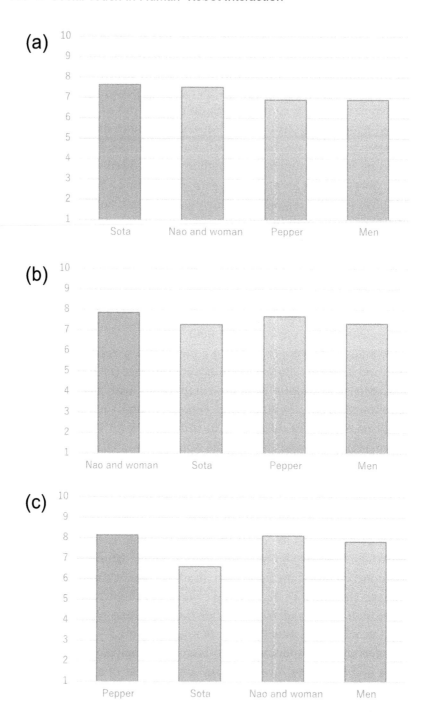

FIGURE 9.3 Comparison with other best combination timings for Sota, Nao, and Pepper: Naturalness of (a) Sota, (b) Nao, and (c) Pepper. The vertical axis showed the questionnaire results.

(mean: 7.52, CI: 7.15–8.16) within its data ($p<0.001$). In addition, Nao's/woman's best combination (mean: 8.15, CI: 7.78–8.61) was non-inferior to Pepper's best combination (mean: 8.11, CI: 7.74–8.57) within its data ($p<0.001$). Therefore, using Nao's/woman's best combination is not the best choice for different robots, although it is a better choice for reaction times.

9.3.3 Limitations

This study has several limitations. First, we only used three specific robots. Different kinds of robots, such as robot-like [21–23], creature-like [24], baby-like [25], animal-like [26], and androids with quite human-like appearances [27,28], would undoubtedly provide different knowledge about reaction times. Moreover, we used relatively simple reaction behaviors, i.e., turning around. If the robots used different reaction behaviors, such as a voice reaction, the appropriate reaction timings would also change. Similar to the limitations of the robot model, we only used videos of two human models; if we prepared children/seniors videos, the survey would provide richer knowledge. From another perspective, appropriate reaction times differ when robots anticipate touch behaviors. For example, when we recognize a situation in which we might be touched, we usually react before actually allowing ourselves to be touched as a pre-touch reaction [29–31]. Investigating natural reaction times before being touched is another future work. Although several limitations exist, we believe that our study provides basic knowledge related to reaction behavior design.

9.4 CONCLUSIONS

Reaction time design is essential for achieving natural human–robot interaction, particularly in social touch contexts. Although various studies have investigated appropriate reaction times for designing robots' behaviors, reaction times during touch interactions have received less focus. Therefore, this study used a web survey to investigate the appropriate reaction times for robots when they are touched. We prepared video stimuli with three robot models and two human models, and we focused on reaction time design by separating it into two phases: a before-reaction time (the length between a touch and the start of a reaction behavior) and an after-reaction time (the lengths of the reaction behavior). We employed 200-ms time resolutions to investigate these reaction time effects between zero and one second.

We individually identified the appropriate reaction time for each robot in the web survey. The experiment results showed a trend in which the body sizes of the models are related to their appropriate reaction times, e.g., a slow before-reaction time and a fast after-reaction time might be better for relatively small robots, whereas a fast before-reaction time and a slow after-reaction time might be better for relatively large robots or human models. We also investigated whether the best combinations of each model are applicable for different robots and found that 400-ms before-reaction times and 600-ms after-reaction times are better for several robots.

ACKNOWLEDGMENT

This work was supported by JST CREST Grant Number JPMJCR18A1, Japan.

REFERENCES

[1] T. Goodman, and R. Spence, "The effect of system response time on interactive computer aided problem solving," In *ACM SIGGRAPH Computer Graphics*, North Carolina, United States, pp. 100–104, 1978.

[2] J. L. Guynes, "Impact of system response time on state anxiety," *Communications of the ACM*, vol. 31, no. 3, pp. 342–347, 1988.

[3] R. B. Miller, "Response time in man-computer conversational transactions," In *Proceedings of the December 9–11, 1968, Fall Joint Computer Conference, Part I,* New York, NY, United States, pp. 267–277, 1968.

[4] T. Kanda, M. Kamasima, M. Imai, T. Ono, D. Sakamoto, H. Ishiguro, and Y. Anzai, "A humanoid robot that pretends to listen to route guidance from a human," *Autonomous Robots*, vol. 22, no. 1, p. 87, 2007.

[5] A. Y. Alhaddad, J.-J. Cabibihan, and A. Bonarini, "Influence of reaction time in the emotional response of a companion robot to a child's aggressive interaction," *International Journal of Social Robotics, vol.* 12, no. 11, p. 1279–1291, 2020.

[6] M. R. Frederiksen, and K. Stoy, "On the causality between affective impact and coordinated human-robot reactions," In *2020 29th IEEE International Conference on Robot and Human Interactive Communication (RO-MAN)*, Naples, Italy, p. 488–494, 2020.

[7] M. Shiomi, T. Minato, and H. Ishiguro, "Subtle reaction and response time effects in human-robot touch interaction," In *International Conference on Social Robotics*, Tsukuba, Japan, p. 242–251, 2017.

[8] M. Shiomi, K. Shatani, T. Minato, and H. Ishiguro, "Does a robot's subtle pause in reaction time to people's touch contribute to positive influences?," In *2018 27th IEEE International Symposium on Robot and Human Interactive Communication (RO-MAN)*, Nanjing and Tai'an, China, pp. 364–369, 2018.

[9] M. Yamamoto, and T. Watanabe, "Time delay effects of utterance to communicative actions on greeting interaction by using a voice-driven embodied interaction system," In *Proceedings of the 2003 IEEE International Symposium on Computational Intelligence in Robotics and Automation,* Kobe, Japan, pp. 217–222, 2003.

[10] T. Shiwa, T. Kanda, M. Imai, H. Ishiguro, and N. Hagita, "How quickly should a communication robot respond? Delaying strategies and habituation effects," *International Journal of Social Robotics*, vol. 1, no. 2, pp. 141–155, 2009.

[11] T. Kanda, M. Shiomi, Z. Miyashita, H. Ishiguro, and N. Hagita, "A communication robot in a shopping mall," *IEEE Transactions on Robotics*, vol. 26, no. 5, pp. 897–913, 2010.

[12] M. Shiomi, T. Kanda, I. Howley, K. Hayashi, and N. Hagita, "Can a social robot stimulate science curiosity in classrooms?," *International Journal of Social Robotics*, vol. 7, no. 5, pp. 641–652, 2015.

[13] D. F. Glas, K. Wada, M. Shiomi, T. Kanda, H. Ishiguro, and N. Hagita, "Personal greetings: Personalizing robot utterances based on novelty of observed behavior," *International Journal of Social Robotics*, vol. 9, no. 2, pp. 181–198, 2017.

[14] A. Kubota, M. Kimoto, T. Iio, K. Shimohara, and M. Shiomi, "From when to when: Evaluating naturalness of reaction time via viewing turn around behaviors," *Applied Sciences*, vol. 11, no. 23, p. 11424, 2021.

[15] F. Galton, "Exhibition of instruments (1) for testing perception of differences of tint, and (2) for determining reaction-time," *The Journal of the Anthropological Institute of Great Britain and Ireland*, vol. 19, pp. 27–29, 1890.

[16] A. T. Welford, *Reaction Times*. London: Academic Press, 1980.

[17] J. Brebner, "Reaction time in personality theory," In Reaction Times, A. T. Welford (ed.), New York: Academic Press. pp. 309–320, 1980.

[18] E. S. Robinson, "Work of the integrated organism," In Handbook of General Experimental Psychology, C. Murchison (ed.), Worcester, MA: Clark University Press. pp. 571–650, 1934.

[19] P. Lele, D. Sinclair, and G. Weddell, "The reaction time to touch," *The Journal of Physiology*, vol. 123, no. 1, pp. 187–203, 1954.

[20] J. S. Downs, M. B. Holbrook, S. Sheng, and L. F. Cranor, "Are your participants gaming the system? screening mechanical turk workers," In *Proceedings of the SIGCHI Conference on Human Factors in Computing Systems,* Atlanta, Georgia, USA, pp. 2399–2402, 2010.

[21] T. Kanda, H. Ishiguro, T. Ono, M. Imai, and R. Nakatsu, "Development and evaluation of an interactive humanoid robot "Robovie"," In *Proceedings 2002 IEEE International Conference on Robotics and Automation (Cat. No.02CH37292),* Washington, DC, United States. vol. 2, pp. 1848–1855, 2002.

[22] R. Matsumura, M. Shiomi, K. Nakagawa, K. Shinozawa, and T. Miyashita, "A desktop-sized communication robot: "Robovie-mr2"," *Journal of Robotics and Mechatronics*, vol. 28, no. 1, pp. 107–108, 2016.

[23] R. Matsumura, and M. Shiomi, "An animation character robot that increases sales," *Applied Sciences*, vol. 12, no. 3, p. 1724, 2022.

[24] N. Yoshida, S. Yonemura, M. Emoto, K. Kawai, N. Numaguchi, H. Nakazato, S. Otsubo, M. Takada, and K. Hayashi, "Production of character animation in a home robot: A case study of lovot," *International Journal of Social Robotics*, vol. 14, no. 1, pp. 39–54, 2022.

[25] H. Sumioka, N. Yamato, M. Shiomi, and H. Ishiguro, "A minimal design of a human infant presence: A case study toward interactive doll therapy for older adults with dementia," *Frontiers in Robotics and AI*, vol. 8, p. 633378, 2021.

[26] R. Yu, E. Hui, J. Lee, D. Poon, A. Ng, K. Sit, K. Ip, F. Yeung, M. Wong, and T. Shibata, "Use of a therapeutic, socially assistive pet robot (PARO) in improving mood and stimulating social interaction and communication for people with dementia: Study protocol for a randomized controlled trial," *JMIR Research Protocols*, vol. 4, no. 2, p. e4189. 2015.

[27] D. F. Glas, T. Minato, C. T. Ishi, T. Kawahara, and H. Ishiguro, "Erica: The erato intelligent conversational android," In *2016 25th IEEE International Symposium on Robot and Human Interactive Communication (RO-MAN)*, New York, NY, United States, pp. 22–29, 2016.

[28] M. Shiomi, H. Sumioka, K. Sakai, T. Funayama, and T. Minato, "SŌTO: An android platform with a masculine appearance for social touch interaction," In *Companion of the 2020 ACM/IEEE International Conference on Human-Robot Interaction*, Cambridge, United Kingdom, pp. 447–449, 2020.

[29] M. Shiomi, K. Shatani, T. Minato, and H. Ishiguro, "How should a robot react before people's touch?: Modeling a pre-touch reaction distance for a robot's face," *IEEE Robotics and Automation Letters*, vol. 3, no. 4, pp. 3773–3780, 2018.

[30] D. A. C. Mejía, H. Sumioka, H. Ishiguro, and M. Shiomi, "Modeling a pre-touch reaction distance around socially touchable upper body parts of a robot," *Applied Sciences*, vol. 11, no. 16, p. 7307, 2021.

[31] D. A. C. Mejía, A. Saito, M. Kimoto, T. Iio, K. Shimohara, H. Sumioka, H. Ishiguro, and M. Shiomi, "Modeling of pre-touch reaction distance for faces in a virtual environment," *Journal of Information Processing*, vol. 29, pp. 657–666, 2021.

CHAPTER 10

Communication Cues Effects in Human–Robot Interaction

Takahiro Hirano, Mitsuhiko Kimoto,
Takamasa Iio, Katsunori Shimohara,
and Masahiro Shiomi

*Advanced Telecommunications Research
Institute International, Kyoto, Japan*

Doshisha University, Kyoto, Japan

10.1 INTRODUCTION

The ability for robots to have a physical presence allows them to interact with people through touch, similar to how humans interact with each other. Such interaction, known as haptic interaction, is a promising research area in the field of human–robot interaction, much like the field of human–human interaction. Such haptic interactions, which have been extensively studied in human science literature, provide both mental and physical benefits [1–6]. Human science literature has reported that haptic interactions positively change people's behavior and support various efforts [7–13]. Previous research has found that physical robots can impact interactions with people more strongly or differently than computer-based agents [14–17]. Building on these findings, researchers have explored the potential positive effects of haptic interaction with

DOI: 10.1201/9781003384274-14

social robots in mental therapy [18], increased motivation [19], and changing attitudes through touching [20–22].

For more natural touch interactions, communication cues such as gaze behavior are essential during touch. Past studies have thoroughly studied the effects of communication cues in non-haptic interaction, including approaching situations [23], encounter situations [24,25], object-handling [26,27], and conversations [28–31]. These studies suggest that gaze behavior design should be altered based on context to create natural interactions. For example, Gharbi et al. [26] emphasized that a robot must express its intention by looking at an object when it hands the object to somebody. Unfortunately, the effects of gaze behavior in haptic interaction have not been sufficiently studied.

Similar to gaze behaviors, touch styles are another essential factor for natural interaction. Previous research has identified three categories of touch styles in haptic interactions with social robots: touching a robot, being touched by a robot, and mutual touch (where a person touches a robot's hand and the robot touches the person's hand) [19,21,22]. These studies examined the effects of touch style on people's impressions of robots, but they did not compare all three touch styles together. As a result, it remains unclear which touch style is most conducive to natural touch interaction.

Based on these considerations, which gaze behavior and touch style combinations are more effective remains unknown. Investigating the effects of each combination among gaze behaviors and touch styles will contribute to creating design guidelines for human–robot touch interaction. Therefore, in this study, we experimentally investigated the effects of these two communication cues with a social robot (Figure 10.1). Note that this chapter is modified based on our previous work [32], edited to be comprehensive and fit with the context of this book.

10.2 MATERIALS AND METHODS OF EXPERIMENT

10.2.1 Robot

We used a humanoid robot, Pepper, developed by Softbank Robotics. The robot has a sufficient number of DOFs as well as five fingers on each hand that provide sufficient capability for human-like touch behavior. Pepper is 121 cm tall. In this study, we conducted an experiment using the Wizard-of-Oz method to accurately control the timing of its behaviors.

10.2.2 Gaze Behavior

Following previous studies that investigated the effectiveness of gaze behaviors in human–robot interaction, we prepared two different gaze

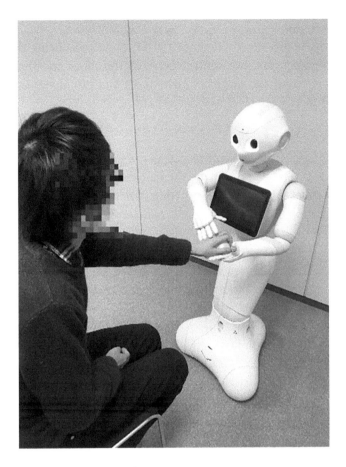

FIGURE 10.1   Pepper robot touches a person.

cues: face-only and face-hand-face. The former is based on the importance of eye contact in human–robot interaction. For example, past studies reported how eye contact behaviors contribute to acceptable interaction for social robots [23,24,30].

Face-hand-face is based on hand-over situations because they offer similar characteristics to a touch situation in the context of the physical distance between a robot and people. A past study [26] investigated the effects of various gaze behaviors in hand-over situations and reported that their participants preferred a gaze behavior in which the robot looked at the object and then at the target person instead of a gaze behavior in which only the target person was looked at, i.e., *face-only* gaze behavior. Based on these considerations, we also prepared a *face-hand-face* gaze behavior in which the robot looks at the participant, its hand, and then the participant again. The following are the details of each behavior.

10.2.2.1 Face-Only

The robot maintains eye contact with a target person during the touch behaviors. For accurate eye contact between the robot and the participants, we measured the height of their faces and calculated the angles toward their faces from the robot beforehand.

10.2.2.2 Face-Hand-Face

The robot changes its face direction due to the touch phases. First, it keeps eye contact with a target person and then looks at the human target's hand. Finally, the robot re-establishes eye contact with the person. The gazing behavior durations and their timing were determined a past work [26].

10.2.3 Touch Style

We prepared three different touch styles based on past studies [19–21]: *touching a robot, being touched by a robot,* and *mutual touching*. The following are the details of each style.

10.2.3.1 Touch-to-Robot

In this style, the robot requests the participants to touch its left hand without touching the participants: "please touch my left hand." Then, it raises its left arm to its chest. To create uniform touch feelings among the other conditions, we placed a plastic yellow ball as a reference point in its left hand.

10.2.3.2 Touched-by-Robot

We set a stand with the plastic yellow ball near the participants on which they could set their hand. In the beginning, the robot says, "please put your hand on the stand" to request the participants to touch the stand. Then, the robot raises its right hand to touch the participant's hand.

We designed the robot's touching behaviors as a stroking on the participants' hands at 5 cm/s by following a past study that described a comfortable touching speed [33]. We prepared a fixed stroking behavior for the robot based on the typical size of the human hand.

10.2.3.3 Mutual Touching

This style combines the touch-to-robot and touched-by-robot conditions. First, the robot asks the participants to touch its left hand, which the robot raises to its chest, similar to the touch-to-robot condition. After its left hand is touched by a participant, the robot reciprocates by raising its right hand to touch the participant's left hand. To create the same feeling of being touched for the participants, we used the same touching motions from the touch-to-robot and touched-by-robot conditions.

10.3 EXPERIMENT

10.3.1 Conditions

As described above, this study investigates two factors: gaze behaviors: two conditions; touch styles: three conditions. We employed a within-participant design, i.e., where all the participants experienced six different touch interactions (2×3) with the robot.

10.3.1.1 Participants

Fourteen women and 14 men participated in our experiment. Their average age was 36.4 and their standard deviation (*SD*) was 9.39.

10.3.1.2 Procedure

First, we explained the purposes and procedures of our experiment. This research was approved by our institution's ethics committee for studies involving human participants. Written informed consent was obtained from every participant. Only those who provided consent joined it.

The participants sat in a chair in front of the robot. The robot greets them and chats briefly with them after requesting a touch from the participants. Because we prepared six conditions, we also prepared six different chat contents to avoid biased impressions. The order of the conditions and the chat contents were counterbalanced. After being touched by the robot in each condition, the participants filled out questionnaires.

10.3.1.3 Measurements

Our questionnaires had the following two items: the *feeling of comfort* of the touch interaction and the robot's *perceived friendliness*. Both were evaluated on a 1-to-7 point scale, where 1 is the most negative and 7 is the most positive.

10.4 RESULTS

Due to the large number of combinations in this study, we only describe the statistical analysis results in this subsection. We incorporated gender factors in the analysis because past studies identified a relationship between touch effects and gender [4,34].

10.4.1 Feelings of Comfort of Touch Interactions

Figures 10.2 and 10.3 show the questionnaire results about their feelings of comfort during the touch interactions. We conducted a three-factor mixed ANOVA for each scale on gaze, touch, and gender and identified significant main effects in the gaze factor ($F(1, 26)=5.253$, $p=0.030$, partial $\eta^2=0.168$), the touch factor ($F(2, 52)=5.706$, $p=0.006$, partial $\eta^2=0.180$), and the simple interaction effect between touch and gender ($F(2, 52)=4.114$, $p=0.022$, partial $\eta^2=0.137$). Multiple comparisons with the Bonferroni method of the simple main effects of touch in men were significant in *touch-to-robot > touched-by-robot* ($p=0.001$) and *touch-to-robot > mutual touching* ($p=0.006$). Multiple comparisons with the Bonferroni method revealed significant differences in the simple main effect of touch in *touched-by-robot* (*women > men*, ($p=0.021$)).

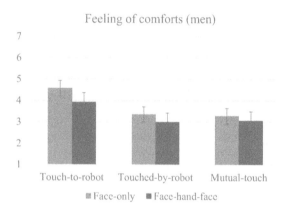

FIGURE 10.2   Feelings of comfort of robot's touch (men).

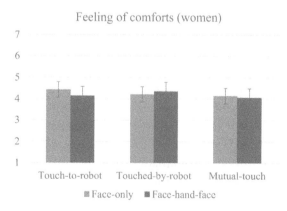

FIGURE 10.3   Feelings of comfort of robot's touch (women).

Next, we summarize the effects of each factor based on the above descriptions. For gaze behavior, participants felt significantly more comfortable in the *face-only* gaze behavior than in the *face-hand-face* gaze behavior ($p<0.05$). Concerning touch style and gender effect, the men felt significantly more comfort in the *touch-to-robot* style than in the other touch styles ($p<0.05$), although women did not show any significant differences.

10.4.2 Perceived Friendliness of Robot

Figures 10.4 and 10.5 show the questionnaire results of perceived friendliness. We conducted a three-factor mixed ANOVA for each scale on gaze, touch, and gender and identified significant main effects in the touch factor (F (2, 52)=3.599, $p=0.034$, partial $\eta^2=0.122$), the gender factor (F (1, 26)=5.484, $p=0.027$, partial $\eta^2=0.174$), the simple interaction effect between gaze and gender (F (1, 26)=4.457, $p=0.045$, partial $\eta^2=0.146$), and the simple interaction effect between touch and gender (F (2, 52)=3.534, $p=0.036$, partial $\eta^2=0.120$). Multiple comparisons with the Bonferroni method revealed significant differences in the simple main effects of gender in *face-hand-face* (women > men ($p=0.013$)). Multiple comparisons with the Bonferroni method revealed significant differences in the simple main effects of touch in *men* (*touch-to-robot* > *touched-by-robot* ($p = 0.003$). Multiple comparisons with the Bonferroni method revealed significant differences in the simple main effects of gender in *touched-by-robot* (women > men ($p=0.007$)) and *mutual touching* (women > men ($p = 0.034$)).

We next summarize the effects of each factor from the above descriptions. None of the gaze behaviors significantly changed the friendliness perceived by the participants; note that women felt significantly higher

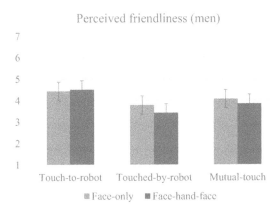

FIGURE 10.4   Perceived friendliness (men).

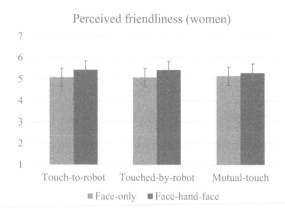

FIGURE 10.5   Perceived friendliness (women).

friendliness in the *face-hand-face* gaze behavior than men ($p<0.05$). For the touch style and the gender effect, men felt significantly higher friendliness in the *touch-to-robot* style than in the *touched-by-robot* style ($p<0.05$), but women did not show any significant differences. Women felt significantly higher friendliness in the *touched-by-robot* and *mutual touching* styles than the men ($p<0.05$).

10.5 DISCUSSION

10.5.1 Design Implications

The experiment results show the importance of maintaining eye contact in touch interaction, similar to conversational interaction [28–31]. On the other hand, this result suggests a contradictory phenomenon compared to handing-over interactions in which people preferred that the robot looked at an object [26]. Since we assume that the purpose of handing interactions is to give a specific item to another, sharing attention by looking at the object is probably preferred. Based on these considerations, preferred gaze behaviors seem to vary based on the interaction between humans and robots.

The experiment results also identified a relatively complex phenomenon about preferred touch styles, because some past studies reported similar results (i.e., people preferred the *touch-to-robot* style more than the *touched-by-robot* style) [20,21], and other past studies reported opposite results (i.e., people preferred the *mutual touching* style over the *touched-by-robot* and *touch-to-robot* styles) [19,35]. These impression changes might reflect the perceived impressions toward the robots and the physical feeling of their hands. In past studies, which reported the positive effects of the *mutual touching* style in human–robot touch interaction, the robots were equipped with soft and comfortable materials on their bodies. On the other hand, other studies, which described the negative effects of the *touched-by-robot* style, including this chapter, put a non-soft material on the bodies of their robots. Therefore, investigating the relationships among the sense of the robot's hand, touch styles, and the perceived impressions is an interesting future work.

10.5.2 Limitations

This study has several limitations since it used a specific robot and a restricted context. As described above, investigating touch style effects with different robots is critical to gather more general knowledge in human–robot touch interactions. Related to this topic, using different robots with various appearances, including robot-like [36–38], creature-like [39], baby-like [40], animal-like [18], and androids with quite human-like appearances [41,42] will undoubtedly influence the perceived impressions. For example, if we use cuter, animal-like pet robots, participants will most likely have more positive experiences to touches from them. On the other hand, if we use more realistic human-like robots, participants might feel more negative toward being touched by them, particularly if their relationships are not close.

Another limitation is our touch behavior itself. In this study, the robot only makes a simple contact with the participants. Past studies reported that different touch characteristics are useful for conveying different emotions [43–46], and whole-body touch interactions such as hugging can effectively improve people's perceptions [47–50]. Various touch styles during such complex touch interactions produce different feelings.

10.6 CONCLUSIONS

Gaze behaviors and touch styles are essential communication cues in human–robot touch interaction. Understanding the effects of various combinations of such communication cues is critical for designing robot's touch behavior that can achieve acceptable and natural touch interaction. Therefore, based on related studies, we prepared two gaze behaviors and three touch styles and experimentally investigated how such communication cues change the perceptions of participants who physically interacted with a robot.

The results showed that participants preferred a robot's gaze behavior that maintains eye contact more than a gaze behavior that combines eye contact and looking at the robot's hand. The participants also preferred a touch style that actively touches the robot more than other touch styles, including being touched by a robot. Although some results showed phenomena that contradicted past studies, the knowledge from our study contributes to understanding how multi-modal touch interaction changes people's perceptions.

ACKNOWLEDGMENT

This work was supported by JST CREST Grant Number JPMJCR18A1, Japan.

REFERENCES

[1] K. M. Grewen, B. J. Anderson, S. S. Girdler, and K. C. Light, "Warm partner contact is related to lower cardiovascular reactivity," *Behavioral Medicine*, vol. 29, no. 3, pp. 123–130, 2003.

[2] S. Cohen, D. Janicki-Deverts, R. B. Turner, and W. J. Doyle, "Does hugging provide stress-buffering social support? A study of susceptibility to upper respiratory infection and illness," *Psychological Science*, vol. 26, no. 2, pp. 135–147, 2015.

[3] B. K. Jakubiak, and B. C. Feeney, "Keep in touch: The effects of imagined touch support on stress and exploration," *Journal of Experimental Social Psychology*, vol. 65, pp. 59–67, 2016.

[4] A. Gallace, and C. Spence, "The science of interpersonal touch: An overview," *Neuroscience & Biobehavioral Reviews*, vol. 34, no. 2, pp. 246–259, 2010.

[5] K. C. Light, K. M. Grewen, and J. A. Amico, "More frequent partner hugs and higher oxytocin levels are linked to lower blood pressure and heart rate in premenopausal women," *Biological Psychology*, vol. 69, no. 1, pp. 5–21, 2005.

[6] T. Field, "Touch for socioemotional and physical well-being: A review," *Developmental Review*, vol. 30, no. 4, pp. 367–383, 2010.

[7] J. D. Fisher, M. Rytting, and R. Heslin, "Hands touching hands: Affective and evaluative effects of an interpersonal touch," *Sociometry*, vol. 39, no. 4, pp. 416–421, 1976.

[8] D. E. Smith, J. A. Gier, and F. N. Willis, "Interpersonal touch and compliance with a marketing request," *Basic and Applied Social Psychology*, vol. 3, no. 1, pp. 35–38, 1982.

[9] J. K. Burgoon, D. B. Buller, J. L. Hale, and M. A. Turck, "Relational messages associated with nonverbal behaviors," *Human Communication Research*, vol. 10, no. 3, pp. 351–378, 1984.

[10] J. Hornik, "Effects of physical contact on customers' shopping time and behavior," *Marketing Letters*, vol. 3, no. 1, pp. 49–55, 1992.

[11] N. Guéguen, "Touch, awareness of touch, and compliance with a request," *Perceptual and Motor Skills*, vol. 95, no. 2, pp. 355–360, 2002.

[12] N. Guéguen, and C. Jacob, "The effect of touch on tipping: An evaluation in a French bar," *International Journal of Hospitality Management*, vol. 24, no. 2, pp. 295–299, 2005.

[13] N. Guéguen, C. Jacob, and G. Boulbry, "The effect of touch on compliance with a restaurant's employee suggestion," *International Journal of Hospitality Management*, vol. 26, no. 4, pp. 1019–1023, 2007.

[14] J. Li, "The benefit of being physically present: A survey of experimental works comparing copresent robots, telepresent robots and virtual agents," *International Journal of Human-Computer Studies*, vol. 77, pp. 23–37, 2015.

[15] W. A. Bainbridge, J. Hart, E. S. Kim, and B. Scassellati, "The effect of presence on human-robot interaction," In *RO-MAN 2008-The 17th IEEE International Symposium on Robot and Human Interactive Communication, Munich, Germany*, pp. 701–706, 2008.

[16] A. Powers, S. Kiesler, S. Fussell, and C. Torrey, "Comparing a computer agent with a humanoid robot," In *2007 2nd ACM/IEEE International Conference on Human-Robot Interaction (HRI), Washington DC, United States*, pp. 145–152, 2007.

[17] K. Shinozawa, F. Naya, J. Yamato, and K. Kogure, "Differences in effect of robot and screen agent recommendations on human decision-making," *International Journal of Human-Computer Studies*, vol. 62, no. 2, pp. 267–279, 2005.

[18] R. Yu, E. Hui, J. Lee, D. Poon, A. Ng, K. Sit, K. Ip, F. Yeung, M. Wong, and T. Shibata, "Use of a therapeutic, socially assistive pet robot (PARO) in improving mood and stimulating social interaction and communication for people with dementia: Study protocol for a randomized controlled trial," *JMIR Research Protocols*, vol. 4, no. 2, p. e4189, 2015.

[19] M. Shiomi, K. Nakagawa, K. Shinozawa, R. Matsumura, H. Ishiguro, and N. Hagita, "Does a robot's touch encourage human effort?," *International Journal of Social Robotics*, vol. 9, pp. 5–15, 2017.

[20] H. Cramer, N. Kemper, A. Amin, B. Wielinga, and V. Evers, "'Give me a hug': The effects of touch and autonomy on people's responses to embodied social agents," *Computer Animation and Virtual Worlds*, vol. 20, no. 2–3, pp. 437–445, 2009.

[21] H. Cramer, N. Kemper, A. Amin, and V. Evers, "Touched by robots: Effects of physical contact and robot proactiveness," In *Workshop on the Reign of Catz and Dogz in CHI, Boston, Massachusetts, United States*, 2009.

[22] T. L. Chen, C.-H. A. King, A. L. Thomaz, and C. C. Kemp, "An investigation of responses to robot-initiated touch in a nursing context," *International Journal of Social Robotics*, vol. 6, no. 1, pp. 141–161, 2013.

[23] S. Satake, T. Kanda, D. F. Glas, M. Imai, H. Ishiguro, and N. Hagita, "A robot that approaches pedestrians," *IEEE Transactions on Robotics*, vol. 29, no. 2, pp. 508–524, 2013.

[24] K. Hayashi, M. Shiomi, T. Kanda, N. Hagita, and A. I. Robotics, "Friendly patrolling: A model of natural encounters," In *Proceedings of the RSS, Sydney, NSW, Australia*, pp. 121, 2012.

[25] C. Shi, M. Shiomi, T. Kanda, H. Ishiguro, and N. Hagita, "Measuring communication participation to initiate conversation in human-robot interaction," *International Journal of Social Robotics*, vol. 7, no. 5, pp. 889–910, 2015.

[26] M. Gharbi, P. V. Paubel, A. Clodic, O. Carreras, R. Alami, and J. M. Cellier, "Toward a better understanding of the communication cues involved in a human-robot object transfer," In *2015 24th IEEE International Symposium on Robot and Human Interactive Communication (RO-MAN), Kobe, Japan*, pp. 319–324, 2015.

[27] C. Shi, M. Shiomi, C. Smith, T. Kanda, and H. Ishiguro, "A model of distributional handing interaction for a mobile robot," In *Robotics: Science and Systems*, Berlin, Germany, pp., 2013.

[28] C. Breazeal, C. D. Kidd, A. L. Thomaz, G. Hoffman, and M. Berlin, "Effects of nonverbal communication on efficiency and robustness in human-robot teamwork," In *2005 IEEE/RSJ International Conference on Intelligent Robots and Systems, Edmonton, Alberta, Canada*, pp. 708–713, 2005.

[29] Y. Kuno, K. Sadazuka, M. Kawashima, K. Yamazaki, A. Yamazaki, and H. Kuzuoka, "Museum guide robot based on sociological interaction analysis," In *Proceedings of the SIGCHI Conference on Human Factors in Computing Systems, San Jose, California, USA*, pp. 1191–1194, 2007.

[30] B. Mutlu, T. Shiwa, T. Kanda, H. Ishiguro, and N. Hagita, "Footing in human-robot conversations: How robots might shape participant roles using gaze cues," In *Proceedings of the 4th ACM/IEEE International Conference on Human Robot Interaction*, California, United States, pp. 61–68, 2009.

[31] M. Shiomi, K. Nakagawa, and N. Hagita, "Design of a gaze behavior at a small mistake moment for a robot," *Interaction Studies*, vol. 14, no. 3, pp. 317–328, 2013.

[32] T. Hirano, M. Shiomi, T. Iio, M. Kimoto, I. Tanev, K. Shimohara, and N. Hagita, "How do communication cues change impressions of human-robot touch interaction?," *International Journal of Social Robotics*, vol. 10, no. 1, pp. 21–31, 2018.

[33] G. K. Essick, A. James, and F. P. McGlone, "Psychophysical assessment of the affective components of non-painful touch," *Neuroreport*, vol. 10, no. 10, pp. 2083–2087, 1999.

[34] B. A. Martin, "A stranger's touch: Effects of accidental interpersonal touch on consumer evaluations and shopping time," *Journal of Consumer Research*, vol. 39, no. 1, pp. 174–184, 2012.

[35] H. Fukuda, M. Shiomi, K. Nakagawa, and K. Ueda, "'Midas touch' in human-robot interaction: Evidence from event-related potentials during the ultimatum game," In *2012 7th ACM/IEEE International Conference on Human-Robot Interaction (HRI)*, Boston, MA, United States, pp. 131–132, 2012.

[36] T. Kanda, H. Ishiguro, T. Ono, M. Imai, and R. Nakatsu, "Development and evaluation of an interactive humanoid robot "Robovie"," In *Proceedings 2002 IEEE International Conference on Robotics and Automation (Cat. No.02CH37292)*, Washington, DC, United States, vol. 2, pp. 1848–1855, 2002.

[37] R. Matsumura, M. Shiomi, K. Nakagawa, K. Shinozawa, and T. Miyashita, "A desktop-sized communication robot: "Robovie-mr2"," *Journal of Robotics and Mechatronics*, vol. 28, no. 1, pp. 107–108, 2016.

[38] R. Matsumura, and M. Shiomi, "An animation character robot that increases sales," *Applied Sciences*, vol. 12, no. 3, p. 1724, 2022.

[39] N. Yoshida, S. Yonemura, M. Emoto, K. Kawai, N. Numaguchi, H. Nakazato, S. Otsubo, M. Takada, and K. Hayashi, "Production of character animation in a home robot: A case study of lovot," *International Journal of Social Robotics*, vol. 14, no. 1, pp. 39–54, 2022.

[40] H. Sumioka, N. Yamato, M. Shiomi, and H. Ishiguro, "A minimal design of a human infant presence: A case study toward interactive doll therapy for older adults with dementia," *Frontiers in Robotics and AI*, vol. 8, p. 633378, 2021.

[41] D. F. Glas, T. Minato, C. T. Ishi, T. Kawahara, and H. Ishiguro, "Erica: The erato intelligent conversational android," In *2016 25th IEEE International Symposium on Robot and Human Interactive Communication (RO-MAN)*, New York, NY, United States, pp. 22–29, 2016.

[42] M. Shiomi, H. Sumioka, K. Sakai, T. Funayama, and T. Minato, "SŌTO: An android platform with a masculine appearance for social touch inter-action," In *Companion of the 2020 ACM/IEEE International Conference on Human-Robot Interaction*, Cambridge, United Kingdom, pp. 447–449, 2020.

[43] X. Zheng, M. Shiomi, T. Minato, and H. Ishiguro, "What kinds of robot's touch will match expressed emotions?," *IEEE Robotics and Automation Letters*, vol. 5, pp. 127–134, 2019.

[44] X. Zheng, M. Shiomi, T. Minato, and H. Ishiguro, "How can robot make people feel intimacy through touch?," *Journal of Robotics and Mechatronics*, vol. 32, no. 1, pp. 51–58, 2019.

[45] X. Zheng, M. Shiomi, T. Minato, and H. Ishiguro, "Modeling the timing and duration of grip behavior to express emotions for a social robot," *IEEE Robotics and Automation Letters*, vol. 6, no. 1, pp. 159–166, 2020.

[46] M. Shiomi, X. Zheng, T. Minato, and H. Ishiguro, "Implementation and evaluation of a grip behavior model to express emotions for an android robot," *Frontiers in Robotics and AI*, vol. 8, p. 755150, 2021.

[47] M. Shiomi, and N. Hagita, "Audio-visual stimuli change not only robot's hug impressions but also its stress-buffering effects," *International Journal of Social Robotics*, vol. 13, pp. 469–476, 2021.

[48] M. Shiomi, A. Nakata, M. Kanbara, and N. Hagita, "Robot reciprocation of hugs increases both interacting times and self-disclosures," *International Journal of Social Robotics*, vol. 13, pp. 353–361, 2020.

[49] Y. Onishi, H. Sumioka, and M. Shiomi, "Increasing torso contact: Comparing human–human relationships and situations," In *International Conference on Social Robotics*, Singapore, Singapore, pp. 616–625, 2021.

[50] A. E. Block, H. Seifi, O. Hilliges, R. Gassert, and K. J. Kuchenbecker, "In the arms of a robot: Designing autonomous hugging robots with intra-hug gestures," *ACM Transactions on Human-Robot Interaction*, vol. 12, pp. 1–49, 2022.

CHAPTER 11

Gaze and Height Design for Acceptable Touch Behaviors

Masahiro Shiomi, Takahiro Hirano,
Mitsuhiko Kimoto, Takamasa Iio,
and Katsunori Shimohara
*Advanced Telecommunications Research
Institute International, Kyoto, Japan*

Doshisha University, Kyoto, Japan

11.1 INTRODUCTION

Robots that possess a physical presence can engage in touch-based interaction with humans, similar to how humans interact with each other. Haptic interaction is an emerging area of study in the field of human–robot interaction, since touch is a key factor in social bonding between humans and provides both mental and physical benefits [1–6]. As a result, human–robot interaction researchers have extensively examined the effects of human–robot touch interaction, which also manifests both mental and physical benefits [7–11].

To achieve acceptable human–robot touch interaction, communication cues play essential roles during touching, e.g., gaze [12–14], voice [15], body movements [16–18], and blinking behavior [19]. Because of the multimodality in human–robot interaction, investigating combinations of communication cues becomes important to understand their effects in touch

DOI: 10.1201/9781003384274-15

129

interaction. For example, Hirano et al. reported how gaze behaviors and touch style during touch interactions influence the perceived impressions of robot-initiated touch [12].

The past studies reported the effectiveness of communication cues and provided guidelines for natural touch interaction. However, two communication cues related to touching behaviors have not yet received enough attention: eye contact height and speech timing. For the former, people who are being touched might regard that action as more polite when the touchers make eye contact from a similar gaze height, compared to different-height eye contact. When people sit, lie on a bed, or interact with children, such situations often occur due to their lower gaze height. Moreover, changing the gaze height to make eye contact shows both a polite attitude and implicitly suggests the timing at which conversations can begin. In human science literature, some studies reported the importance and effectiveness of eye contact before interaction by adjusting the gaze height [20–22].

In human–robot interaction studies, some studies have also reported the effects of a robot's height in conversational settings. One investigated the height of a telepresence robot and reported that a lower height than interacting people is less persuasive [23]. Another study investigated the acceptable height of a conversational robot and concluded that a conversational partner's gaze height under 300 mm is acceptable [24]. Although they reported gaze-height effects in conversational settings, they focused less on touch situations where robots change their gaze height before they touch somebody. Based on these considerations, this study addresses the following research question:

- How does gaze height in eye contact in human–robot touch interaction influence peoples' perceptions?

Concerning speech timing, human science literature has reported that when a person touches another, the toucher usually informs the person being touched of her intention *before* the touching, especially in nursing contexts [20–22]. Another study argued that nurses must obtain permission to touch before actually making contact, adding that patients want an explanation for being touched before contact [25]. These studies suggested the effectiveness of before-touch speech timing. However, a human–robot touch interaction study reported that participants preferred after-touch speech timing, i.e., favoring a robot's explanations after being touched [15]. These contradictory results complicate designing when robots speak while touching a person. Based on these considerations, this study also addresses the following research question:

- How does speech timing in human–robot touch interaction influence human perceptions?

Note that this chapter is modified based on our previous work [26], edited to be comprehensive and fit with the context of this book.

11.2 RELATED WORK

11.2.1 Gaze Behavior for Social Robot

Due to the importance of gazing behavior design for natural interaction with social robots, researchers have investigated the effective gaze functions in broader situations like multi-party conversations [27], information-providing tasks [28], interaction with multiple children [29], and storytelling tasks for toddlers [30]. Gaze behavior design is also important not only for such conversational situations but also for locomotion situations, e.g., approaching people [31,32] and moving around in daily environments [33].

Some studies focused on gaze behavior design in human–robot touch interaction. One study investigated the effects of combinations of communication cues, i.e., gaze targets and touch styles, and described the effectiveness of maintaining eye contact while touching [12]. Other studies reported the effects of a robot's face height that influences its gaze behaviors. One past study reported that a shorter telepresence robot compared to interacting people is less persuasive in a conversational setting [23]. On the other hand, another study reported that a robot with a low face position is preferred as a conversational partner [24].

However, these studies less focused on gaze-height factors during touch interaction. Therefore, it remains unknown how a robot's gaze height influences perceived feelings in touch contexts.

11.2.2 Speech Timing for Social Robot

Similar to gaze behavior design, speech timing is also broadly investigated in human–robot interaction contexts due to its importance. Because it is strongly related to conversational behaviors, researchers investigated appropriate speech timing and related behaviors under conversational tasks, such as route-guidance [34] and filler design in conversations [35,36]. To achieve cognitively understandable speech, a past study investigated appropriate speech rates to decrease cognitive load during conversations [37].

Although speech timing for social robots has been thoroughly investigated in conversational settings, knowledge is limited in touch settings.

A past study of human–robot touch interaction concluded that in a nursing setting, participants preferred a robot's speech after being touched to before being touched [15], although another human–human touch interaction study suggested that conversation before touching is more polite and acceptable in nursing contexts [25]. Therefore, the appropriate speech timing remains unknown for a social robot that actively touches people.

11.3 MATERIALS AND METHODS OF EXPERIMENT

11.3.1 Robot and Sensor

We used a humanoid robot called Pepper, which has 20 degrees of freedom (DOF) and is 121 cm tall. It has enough capabilities for a robot-initiated touch of people by changing its gaze height by bending at its waist. We installed a touch sensor on its right hand to detect physical contact during robot-initiated touches. We used Shokkaku Cube, which can measure the height changes on the top of the cube's soft materials with 100-Hz frequency. The sensor system is connected to a robot's control system by a network and manages the timing at which the robot must stop its motions for safety.

11.3.2 Situation

In this study, we followed a similar touching situation of a past study [15] to investigate the effects of two communication cues (gaze height and speech timing), i.e., participants lying on a bed. Because of the robot's motion capabilities, we adjusted the bed's height to enable the robot to touch the participants. Based on this adjustment, the robot's gaze height is higher than the participants; when it bends down, its gaze height becomes closer to the participants' face height. To provide an identical touching stimuli to the participants as much as possible, we prepared the fixed robot's arm trajectory to touch the left shoulders of the participants.

11.3.3 Gaze-Height Design

We investigated the effects of two different gaze heights during touching: crouching down and looking down. The former is where the robot makes eye contact with participants after crouching down to change its own face height (Figure 11.1). The latter is where the robot makes eye contact without changing its face height (Figure 11.2). We prepared the crouching-down condition since human science literature has reported the positive effects of making eye contact at the same gaze height [20–22]. On the other hand, since human–robot touch interaction studies have focused less on gaze height

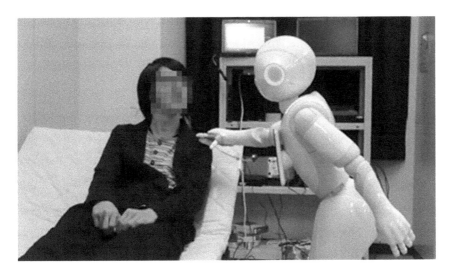

FIGURE 11.1   Robot's touch with crouching-down.

FIGURE 11.2   Robot's touch with looking-down.

during touch interaction, i.e., a robot touching a person without changing its face height, we prepared a looking-down condition as an alternative.

11.3.4 Speech-Timing Design

In this study, we investigated the effects of two different speech timings: before-touch and after-touch. In the former, the robot explains why it is touching before the actual touch is made, i.e., the robot conveys its intention beforehand. The latter is the opposite behavior; the robot gives a reason for touching after touching, i.e., it conveys its intention afterwards. As we explained above, since human science literature suggests that patients prefer a spoken explanation before being touched in a nursing context [20–22,25], we prepared a before-touch condition. On the other hand, since a human–robot touch interaction study suggested that participants preferred a spoken explanation after being touched [15], we prepared an after-touch condition as an alternative.

11.3.5 Procedure

Before starting the experiment, the experimenter described its purpose and asked the participants to fill out consent forms. The experiment procedure was approved by an ethics committee of our institution. Next, the experimenter explained the experiment's procedures and asked the participants to imagine a medical context before they got on the bed. We adjusted its height and the robot's position for its touching behavior.

We employed a within-participant design. All the participants joined four sessions due to combinations of a gaze-height factor (crouching down and looking down) and a speech-timing factor (before-touch and after-touch). We assigned different condition orders to avoid any order effects by considering counterbalances. After experiencing each condition, the participants filled out questionnaires.

11.3.6 Measurements

We evaluated the effects of gaze height and speech timing in human–robot touch interaction in a nursing context with the following three questionnaire items: comfortableness, likeability, and safety. We employed an existing questionnaire item [12] on comfortableness. We also employed existing scales for likeability and safety [38].

11.3.7 Participants

In this study, 32 participants joined the experiment: 16 women and 16 men, whose ages averaged 22.9, SD: 1.69, ranging between 21 and 27. We confirmed that none had any previous experience of touch interaction with Pepper.

11.4 RESULTS AND DISCUSSIONS

11.4.1 Questionnaire Results

We conducted a two-factor repeated measures ANOVA on the conditions, and an analysis of *comfortableness* (Figure 11.3) showed significant differences in the speech-timing factor (F (1, 31) = 18.086, $p < 0.001$, partial

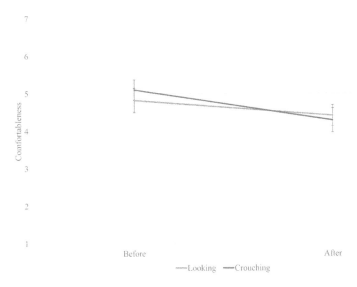

FIGURE 11.3 Questionnaire results about comfortableness.

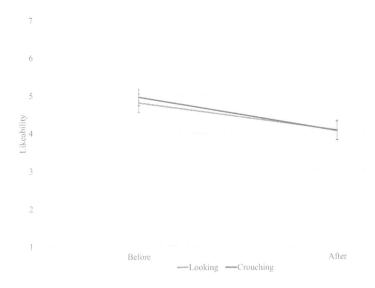

FIGURE 11.4 Questionnaire results about likeability.

$\eta^2=0.368$). No significance was found in the gaze-height factor (F (1, 31)=0.155, $p=0.696$, partial $\eta^2=0.005$) or in the interaction effect (F (1, 31)=1.108, $p=0.301$, partial $\eta^2=0.035$).

We conducted a two-factor repeated measures ANOVA on the conditions, and an analysis of *likeability* (Figure 11.4) showed significant differences in the speech-timing factor (F (1, 31)=26.694, $p<0.001$,

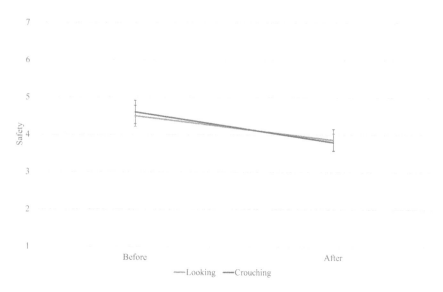

FIGURE 11.5   Questionnaire results about safety.

partial $\eta^2 = 0.463$). No significance was found in the gaze-height factor (F (1, 31) = 0.191, $p = 0.665$, partial $\eta^2 = 0.006$) or in the interaction effect (F (1, 31) = 0.702, $p = 0.409$, partial $\eta^2 = 0.022$).

We conducted a two-factor repeated measures ANOVA on the conditions, and an analysis of *safety* (Figure 11.5) showed significant differences in the speech-timing factor (F (1, 31) = 14.358, $p = 0.001$, partial $\eta^2 = 0.317$). No significance was found in the gaze-height factor (F (1, 31) = 0.015, $p = 0.904$, partial $\eta^2 = 0.001$) or the interaction effect (F (1, 31) = 0.585, $p = 0.450$, partial $\eta^2 = 0.019$).

11.4.2 Design Implications of Speech Timing

Our experiment results on speech-timing effects showed an opposite phenomenon compared to a past study [15]. Although it described the effectiveness of after-touch timing, our study found an advantage of before-touch timing. Several factors might have caused such contradictory results: robot's characteristics, communication cues, and cultural differences. For example, we used a Pepper robot in this study, which has a different appearance and touch feeling compared to the robot in the past study. Moreover, Pepper can make eye contact with the participants. Keeping eye contact, which is an effective communication cue during touch interaction [12], may signal the robot's intention (i.e., touching) to the participants before touching. Therefore, before-touch speech timing seems appropriate for keeping eye contact before touching.

11.4.3 Discussion: Design Implications of Gaze Height

Unlike the speech-timing factor, the gaze-height factor showed no significant effects during touch interaction, although past studies reported the effectiveness of gaze-height design in conversational settings [20–22]. We also expected that several factors might cause such results: robot's eye design, posture, characteristics, communication cues, and cultural differences. Concerning eye design, since Pepper's gazes are designed to make eye contact from any angle, participants may not experience unnatural feelings even though the robot's head is at a different level. The participants in the past study were prone on a flatbed [15], which might have caused a different feeling compared to the past study setting.

11.4.4 Limitations

This study has several limitations since we used a specific robot (Pepper) and a specific touch (touching the left shoulders of the participants). Conducting similar experiments with different kinds of robots, including different robots [39–41] or androids with quite human-like appearances [42,43], would provide more detailed information about the effects of communication cues in human–robot touch interaction. Investigating the effects of different touch behaviors that are designed to convey emotions [44–47] would also provide rich knowledge about the combination effects of affective touches and communication cues.

11.5 CONCLUSIONS

Communication cues in human–robot touch interaction have essential roles for natural and acceptable robot-initiated touches. We investigated two communication cues where a robot touches people in a nursing context: gaze height and speech timing. For the former, to improve eye contact in touch interaction, we compared the effects of a crouching-down behavior at the same level and looking-down to a different level. In terms of the latter, to unveil which timing is better in touch interaction, we compared the before-touch and after-touch effects. We prepared four touching behaviors for Pepper, conducted an experiment with 32 participants, and found that they preferred before-touch speech timing over after-touch speech timing. However, our results did not show any advantages of crouching-down behavior for making the same-level eye contact.

ACKNOWLEDGMENT

This work was supported by JST CREST Grant Number JPMJCR18A1, Japan.

REFERENCES

[1] K. M. Grewen, B. J. Anderson, S. S. Girdler, and K. C. Light, "Warm partner contact is related to lower cardiovascular reactivity," *Behavioral Medicine*, vol. 29, no. 3, pp. 123–130, 2003.

[2] S. Cohen, D. Janicki-Deverts, R. B. Turner, and W. J. Doyle, "Does hugging provide stress-buffering social support? A study of susceptibility to upper respiratory infection and illness," *Psychological Science*, vol. 26, no. 2, pp. 135–147, 2015.

[3] B. K. Jakubiak, and B. C. Feeney, "Keep in touch: The effects of imagined touch support on stress and exploration," *Journal of Experimental Social Psychology*, vol. 65, pp. 59–67, 2016.

[4] A. Gallace, and C. Spence, "The science of interpersonal touch: An overview," *Neuroscience & Biobehavioral Reviews*, vol. 34, no. 2, pp. 246–259, 2010.

[5] K. C. Light, K. M. Grewen, and J. A. Amico, "More frequent partner hugs and higher oxytocin levels are linked to lower blood pressure and heart rate in premenopausal women," *Biological Psychology*, vol. 69, no. 1, pp. 5–21, 2005.

[6] T. Field, "Touch for socioemotional and physical well-being: A review," *Developmental Review*, vol. 30, no. 4, pp. 367–383, 2010.

[7] R. Yu, E. Hui, J. Lee, D. Poon, A. Ng, K. Sit, K. Ip, F. Yeung, M. Wong, and T. Shibata, "Use of a therapeutic, socially assistive pet robot (PARO) in improving mood and stimulating social interaction and communication for people with dementia: Study protocol for a randomized controlled trial," *JMIR Research Protocols*, vol. 4, no. 2, p. e4189, 2015.

[8] M. Shiomi, K. Nakagawa, K. Shinozawa, R. Matsumura, H. Ishiguro, and N. Hagita, "Does a robot's touch encourage human effort?," *International Journal of Social Robotics*, vol. 9, pp. 5–15, 2016.

[9] H. Sumioka, A. Nakae, R. Kanai, and H. Ishiguro, "Huggable communication medium decreases cortisol levels," *Scientific Reports*, vol. 3, p. 3034, 2013.

[10] M. Shiomi, A. Nakata, M. Kanbara, and N. Hagita, "A hug from a robot encourages prosocial behavior," In *2017 26th IEEE International Symposium on Robot and Human Interactive Communication (RO-MAN)*, Lisbon, Portugal, pp. 418–423, 2017.

[11] M. Shiomi, and N. Hagita, "Audio-visual stimuli change not only robot's hug impressions but also its stress-buffering effects," *International Journal of Social Robotics*, vol. 13, pp. 469–476, 2021.

[12] T. Hirano, M. Shiomi, T. Iio, M. Kimoto, I. Tanev, K. Shimohara, and N. Hagita, "How do communication cues change impressions of human-robot touch interaction?," *International Journal of Social Robotics*, vol. 10, no. 1, pp. 21–31, 2018.

[13] Y. Tamura, T. Akashi, and H. Osumi, "Where robot looks is not where person thinks robot looks," *Journal of Advanced Computational Intelligence and Intelligent Informatics*, vol. 21, no. 4, pp. 660–666, 2017.

[14] H. Sumioka, Y. Yoshikawa, and M. Asada, "Learning of joint attention from detecting causality based on transfer entropy," *Journal of Robotics Mechatronics*, vol. 20, no. 3, p. 378, 2008.

[15] T. L. Chen, C.-H. A. King, A. L. Thomaz, and C. C. Kemp, "An investigation of responses to robot-initiated touch in a nursing context," *International Journal of Social Robotics*, vol. 6, no. 1, pp. 141–161, 2013.

[16] T. L. Chen, T. Bhattacharjee, J. M. Beer, L. H. Ting, M. E. Hackney, W. A. Rogers, and C. C. Kemp, "Older adults' acceptance of a robot for partner dance-based exercise," *PLoS One*, vol. 12, no. 10, p. e0182736, 2017.

[17] T. L. Chen, T. Bhattacharjee, J. L. McKay, J. E. Borinski, M. E. Hackney, L. H. Ting, and C. C. Kemp, "Evaluation by expert dancers of a robot that performs partnered stepping via haptic interaction," *PLoS One*, vol. 10, no. 5, p. e0125179, 2015.

[18] K. Kosuge, T. Hayashi, Y. Hirata, and R. Tobiyama, "Dance partner robot-ms dancer," In *Proceedings of the 2003 IEEE/RSJ International Conference on Intelligent Robots and Systems*, Las Vegas, Nevada, United States, pp. 3459–3464, 2003.

[19] K. Funakoshi, K. Kobayashi, M. Nakano, S. Yamada, Y. Kitamura, and H. Tsujino, "Smoothing human-robot speech interactions by using a blinking-light as subtle expression," In *Proceedings of the 10th International Conference on Multimodal Interfaces*, pp. 293–296, 2008.

[20] Z. N. Kain, J. E. MacLaren, C. Hammell, C. Novoa, M. A. Fortier, H. Huszti, and L. Mayes, "Healthcare provider-child-parent communication in the preoperative surgical setting," *Pediatric Anesthesia*, vol. 19, no. 4, pp. 376–384, 2009.

[21] E. B. Wright, C. Holcombe, and P. Salmon, "Doctors' communication of trust, care, and respect in breast cancer: Qualitative study," *British Medical Journal*, vol. 328, no. 7444, p. 864, 2004.

[22] R. F. Brown, and C. L. Bylund, "Communication skills training: Describing a new conceptual model," *Academic Medicine*, vol. 83, no. 1, pp. 37–44, 2008.

[23] I. Rae, L. Takayama, and B. Mutlu, "The influence of height in robot-mediated communication," In *2013 8th ACM/IEEE International Conference on Human-Robot Interaction (HRI)*, Tokyo, Japan, pp. 1–8, 2013.

[24] Y. Hiroi, and A. Ito, "Influence of the height of a robot on comfortableness of verbal interaction," *IAENG International Journal of Computer Science*, vol. 43, no. 4, pp. 447–455, 2016.

[25] C. O'lynn, and L. Krautscheid, "'How should I touch you?': A qualitative study of attitudes on intimate touch in nursing care," *AJN The American Journal of Nursing*, vol. 111, no. 3, pp. 24–31, 2011.

[26] M. Shiomi, T. Hirano, M. Kimoto, T. Iio, and K. Shimohara, "Gaze-height and speech-timing effects on feeling robot-initiated touches," *Journal of Robotics and Mechatronics*, vol. 32, no. 1, pp. 68–75, 2020.

[27] B. Mutlu, T. Shiwa, T. Kanda, H. Ishiguro, and N. Hagita, "Footing in human-robot conversations: How robots might shape participant roles using gaze cues," In *Proceedings of the 4th ACM/IEEE International Conference on Human Robot Interaction*, California, United States, pp. 61–68, 2009.

[28] Y. Kuno, K. Sadazuka, M. Kawashima, K. Yamazaki, A. Yamazaki, and H. Kuzuoka, "Museum guide robot based on sociological interaction analysis," In *Proceedings of the SIGCHI Conference on Human Factors in Computing Systems*, San Jose, California, USA, pp. 1191–1194, 2007.

[29] T. Komatsubara, M. Shiomi, T. Kanda, H. Ishiguro, and N. Hagita, "Can a social robot help children's understanding of science in classrooms?," In *Proceedings of the 2nd International Conference on Human-Agent Interaction*, Tsukuba, Japan, pp. 83–90, 2014.

[30] Y. Tamura, M. Kimoto, M. Shiomi, T. Iio, K. Shimohara, and N. Hagita, "Effects of a listener robot with children in storytelling," In *Proceedings of the 5th International Conference on Human Agent Interaction*, Bielefeld, Germany, pp. 35–43, 2017.

[31] S. Satake, T. Kanda, D. F. Glas, M. Imai, H. Ishiguro, and N. Hagita, "A robot that approaches pedestrians," *IEEE Transactions on Robotics*, vol. 29, no. 2, pp. 508–524, 2013.

[32] Y. Miyaji, and K. Tomiyama, "Implementation approach of affective interaction for caregiver support robot," *Journal of Robotics and Mechatronics*, vol. 25, no. 6, pp. 1060–1069, 2013.

[33] K. Hayashi, M. Shiomi, T. Kanda, N. Hagita, and A. I. Robotics, "Friendly patrolling: A model of natural encounters," In *Proceedings of the RSS*, Sydney, NSW, Australia, p. 121, 2012.

[34] Y. Okuno, T. Kanda, M. Imai, H. Ishiguro, and N. Hagita, "Providing route directions: Design of robot's utterance, gesture, and timing," In *Proceedings of the 4th ACM/IEEE International Conference on Human Robot Interaction*, California, United States, pp. 53–60, 2009.

[35] T. Shiwa, T. Kanda, M. Imai, H. Ishiguro, and N. Hagita, "How quickly should a communication robot respond? Delaying strategies and habituation effects," *International Journal of Social Robotics*, vol. 1, no. 2, pp. 141–155, 2009.

[36] R. Nakanishi, K. Inoue, S. Nakamura, K. Takanashi, and T. Kawahara, "Generating fillers based on dialog act pairs for smooth turn-taking by humanoid robot," In Proc. International Workshop Spoken Dialogue Systems (IWSDS), pp., 2018.

[37] M. Shimada, and T. Kanda, "What is the appropriate speech rate for a communication robot?," *Interaction Studies*, vol. 13, no. 3, pp. 408–435, 2012.

[38] C. Bartneck, D. Kulić, E. Croft, and S. Zoghbi, "Measurement instruments for the anthropomorphism, animacy, likeability, perceived intelligence, and perceived safety of robots," *International Journal of Social Robotics*, vol. 1, no. 1, pp. 71–81, 2009.

[39] T. Kanda, H. Ishiguro, T. Ono, M. Imai, and R. Nakatsu, "Development and evaluation of an interactive humanoid robot "Robovie"," In *Proceedings 2002 IEEE International Conference on Robotics and Automation (Cat. No.02CH37292)*, Washington, DC, United States, vol. 2, pp. 1848–1855, 2002.

[40] R. Matsumura, M. Shiomi, K. Nakagawa, K. Shinozawa, and T. Miyashita, "A desktop-sized communication robot: "Robovie-mr2"," *Journal of Robotics and Mechatronics*, vol. 28, no. 1, pp. 107–108, 2016.

[41] R. Matsumura, and M. Shiomi, "An animation character robot that increases sales," *Applied Sciences*, vol. 12, no. 3, p. 1724, 2022.

[42] D. F. Glas, T. Minato, C. T. Ishi, T. Kawahara, and H. Ishiguro, "Erica: The erato intelligent conversational android," In *2016 25th IEEE International Symposium on Robot and Human Interactive Communication (RO-MAN)*, New York, NY, United States, pp. 22–29, 2016.

[43] M. Shiomi, H. Sumioka, K. Sakai, T. Funayama, and T. Minato, "SŌTO: An android platform with a masculine appearance for social touch interaction," In *Companion of the 2020 ACM/IEEE International Conference on Human-Robot Interaction*, Cambridge, United Kingdom, pp. 447–449, 2020.

[44] X. Zheng, M. Shiomi, T. Minato, and H. Ishiguro, "What kinds of robot's touch will match expressed emotions?," *IEEE Robotics and Automation Letters*, vol. 5, pp. 127–134, 2019.

[45] X. Zheng, M. Shiomi, T. Minato, and H. Ishiguro, "How can robot make people feel intimacy through touch?," *Journal of Robotics and Mechatronics*, vol. 32, no. 1, pp. 51–58, 2019.

[46] X. Zheng, M. Shiomi, T. Minato, and H. Ishiguro, "Modeling the timing and duration of grip behavior to express emotions for a social robot," *IEEE Robotics and Automation Letters*, vol. 6, no. 1, pp. 159–166, 2020.

[47] M. Shiomi, X. Zheng, T. Minato, and H. Ishiguro, "Implementation and evaluation of a grip behavior model to express emotions for an android robot," *Frontiers in Robotics and AI*, vol. 8, p. 755150, 2021.

CHAPTER 12

Designing Touch Characteristics to Express Simple Emotions

Xiqian Zheng and Masahiro Shiomi

Advanced Telecommunications Research Institute International, Kyoto, Japan

Takashi Minato

RIKEN, Saitama, Japan

Hiroshi Ishiguro

Osaka University, Osaka, Japan

12.1 INTRODUCTION

Emotion expressions are crucial for designing behaviors for social robots as their working environments spread throughout society to provide services, including physical/mental health support [1,2], education [3,4], and companionship [5,6]. Robotics researchers have begun to focus on touch behavior design for such purposes [7–10]. For example, several studies investigated how people touch robots when they want to convey emotions to them [11–13]. Other studies investigated the relationship between which body parts are touched and emotion in human–robot interaction [14,15].

Unfortunately, designing touch interaction for conveying emotions from robots to people hasn't received sufficient focus yet. Although a past

DOI: 10.1201/9781003384274-16

FIGURE 12.1 ERICA's hand touch behaviors (left: using her palm, right: using her fingers).

study investigated touch behavior design with non-humanoid robots [16], they did not use social robots. Other studies focused on speed characteristics during touch interaction to achieve CT-optimal touches (around 3–5 cm/s) [17,18], although they focused less on emotional touch interaction design.

Therefore, it remains unknown what characteristics are essential to convey a robot's emotions by touch interaction. People may change their touch characteristics in human–human interaction due to emotions, e.g., touch duration, style, and specific parts. In human–robot interaction, social robots may need to express emotions to be accepted by interacting people, and if so, changing their touch characteristics is also needed.

Based on these considerations, we investigate the influences of touch characteristics on the perceived emotions of interacting people. We focused on three different touch characteristics (length, type, and body part) with two different emotions (happiness and sadness) from two perspectives (arousal and valence). We conducted an experiment with a human-like android robot called ERICA [19] (Figure 12.1). Note that this chapter is modified based on our previous work [20], edited to be comprehensive and fit with the context of this book.

12.2 TARGET EMOTIONS, TOUCH CHARACTERISTICS, AND ROBOT

12.2.1 Target Emotions

We focused on two emotions (happiness and sadness) because happy emotions are typically used in positive emotional expressions and sad emotions denote negative emotional expressions in human–robot interaction [21,22]. These two are bipolar emotions, based on Russell's circumplex model [23]. Investigating the relationships between touch characteristics and these two emotions that have opposite arouse/valence aspects (happy: high arousal and valence, sad: low arousal and valence) will provide rich knowledge about emotional expression design by touch interaction.

12.2.2 Touch Characteristics

Similar to past studies of human–robot touch interaction [24] and considering an acceptable touchable part in human–human interaction [25,26], we designed a robot to touch a participant's hand to convey emotion. In this section, we describe the characteristics of its hand touch. We avoided excessive combinations of characteristics by focusing on those characteristics related to arousal/valence impressions and selected three candidates: length, type, and body part:

Length: Past studies reported different effects of the length of a touch. One concluded that a longer touch is perceived as a negative valence [27], although another showed no effect of a touch's length [24]. Another study claimed that a longer touch duration is related to high-arousal situations [13] in touch interaction from people to a pet-like robot, but another showed no such length of touch effects [24]. Therefore, investigating the effects of a touch's length will provide additional knowledge to understand its effects on emotional touch design. To decide the touch length for our experiment, we conducted a small pilot study in our laboratory and heuristically chose 0.5- and 2-second contact durations for short and long touches.

Type: Previous studies also reported different effects of the type of touch. A couple of studies described the relationships between high arousal/valence and pulses in touch (i.e., tapping) [13,27], while another reported a smaller effect of touch type for arousal/valence [24]. Similar to touch length, investigating its effects will provide additional knowledge. We determined the touch type in our experiment by small pilot study and heuristically determined a lingering time of 50 ms for short-pat touches and 250 ms for long-pat touches.

Part: Although past studies did not investigate the effect of the touching part (e.g., a hand or fingers), they did investigate the relationships between the intensity of the touch stimuli and arousal. For example, one work reported that high intensity showed high arousal [27]. Another study concluded that strong force showed high arousal but low valence [24]. In this study, we changed the size of the area and the total touch pressure to show different intensities, i.e., using a hand and fingers.

12.2.3 Robot

Figure 12.1 shows ERICA [19], which is used in our study. The robot has three DOFs in the torso and ten in each of arms. The robot also has the two DOFs on each of the wrists and the three on the palms, which enables the robot to realize various touch behaviors. The control system can update each actuator target position every 50 ms. Note that the silicon-based skins realize human-like appearance but have different touch feelings than human skin; therefore, we put gloves on the hands to avoid uncomfortable feelings via touch interaction.

We prepared facial expression motions and Japanese voices to express both emotions: "I'm really happy" in happy emotion, and "I'm so sad" in sad emotion. The timing of starting facial expressions and voices is played when the touch behaviors start. We used HOYA text-to-speech software (http://voicetext.jp/) as a speech synthesis function.

12.3 EXPERIMENT

12.3.1 Hypotheses and Predictions

Touch characteristics are essential factors in expressing emotions. People use different types of touches when they convey their emotions [11–13], therefore we thought that a robot also needs to change its touch characteristics to convey its emotions to people via touch interaction. Although past studies showed different effects of each touch characteristic [24], we focused on three touch characteristics (length, type, and part) that might be important to convey happy and sad emotions, which are typically applied emotions in human–robot interaction [21,22,28–30]. In this study, we made the following three predictions:

Prediction 1: When the robot expresses a happy emotion, a short touch will be perceived as stronger and more natural than a long touch. When it expresses a sad emotion, a long touch will be perceived as stronger and more natural than a short one.

Prediction 2: When the robot expresses a happy emotion, a pat-type touch will be perceived as stronger and more natural than a contact-type touch. When it expresses a sad emotion, the contact-type touch will be perceived as stronger and more natural than the pat-type touch.

Prediction 3: When the robot expresses a happy emotion, a finger touch will be perceived as stronger and more natural than a hand touch. When it expresses a sad emotion, a hand touch will be perceived as stronger and more natural than a finger touch.

12.3.2 Participants

In this study, 22 native Japanese (11 women and 11 men whose ages ranged from 19 to 39 and averaged 29.0) joined the experiment. None had experienced any touch interaction with our robot.

12.3.3 Conditions

We prepared four factors: emotion (happy and sad), length (short and long), type (contact and pat), and part (hand and finger). The experiment has a within-participants design. Each participant experienced 16 trials. The order of the conditions was counterbalanced as much as possible.

12.3.4 Procedure

Before the experiment, the participants were given a brief description of its purpose and procedure. This research was approved by our institution's ethics committee for studies involving human participants. Written informed consent was obtained from each one.

The experimenter explained to the participants how the robot expresses emotions, i.e., using touch, a facial expression, and speech. We positioned the participants next to the robot and placed markers on a table to fix their hand positions to ensure that they experienced the identical touch interaction and easily observed the robot's facial expressions. After the experiment started, the robot first randomly selected one of two emotions (happy or sad) without a touch behavior to show its baseline condition. Then, it expressed the selected emotion 16 times with different touch behaviors by following the pre-ordered combinations. After finishing each trial, the robot asked the participants to compare its strength and naturalness to the baseline condition. Then, it again showed the baseline condition with the last emotion and repeated the same procedure.

12.3.5 Measurements

We measured two impressions of conveyed emotions: strength (degree of strength of the perceived emotion through the android's behaviors) and naturalness (degree of naturalness of the touch behavior to express the emotion) on a 1–7 point scale, where 1 indicates the most negative compared to the baseline, 4 indicates the same impression as the baseline, and 7 indicates the most positive compared to the baseline.

12.4 RESULTS

12.4.1 Statistical Analysis about Strength Impressions

We conducted a four-factor mixed ANOVA for the strength impressions and identified the significant main effects in the type factor (F (1, 21)=5.143, p=.034, partial η^2=.374) and in the part factor (F (1, 21)=10.337, p=.004, partial η^2=.330). We also identified simple interaction effects between emotion and length (F (1, 21)=15,717, p=.001, partial η^2=.428) and emotion and type (F (1, 21)=22.066, p=.001, partial η^2=.512). We found no significant differences in the other simple main and interaction effects.

We conducted multiple comparisons with the Bonferroni method of the simple main effects and identified significant differences in both the strength and naturalness impressions. We found significant differences in happy where short>long (p=.022) and sad where long>short (p=.030). We also found significant differences in happy where pat>contact (p=.012) and sad where contact>pat (p=.001). Figures 12.2–12.4 showed the questionnaire results.

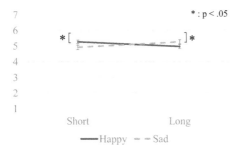

FIGURE 12.2   Average values of strength with touch length.

FIGURE 12.3 Average values of strength with touch type.

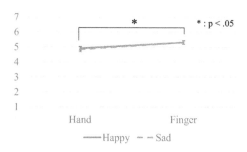

FIGURE 12.4 Average values of strength with touch part.

12.4.2 Statistical Analysis about Naturalness Impressions

We also conducted a four-factor mixed ANOVA for the naturalness impressions and identified significant main effects in the part factor (F (1, 21) = 49.941, p = .001, partial η^2 = .704) and the simple interaction effects between emotion and length (F (1, 21) = 14.384, p = .001, partial η^2 = .407) and emotion and type (F (1, 21) = 28.453, p = .001, partial η^2 = .575). We found no significant differences in other simple main and interaction effects. Figures 12.5–12.7 showed the questionnaire results.

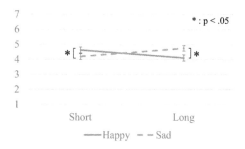

FIGURE 12.5 Average values of naturalness with touch length.

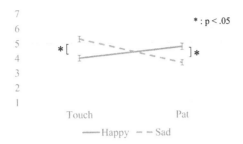

FIGURE 12.6 Average values of naturalness with touch type.

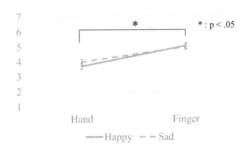

FIGURE 12.7 Average values of naturalness with touch part.

We conducted multiple comparisons with the Bonferroni method of the simple main effects and identified significant differences in both the strength and naturalness impressions. We found significant differences in happy where short>long ($p=.008$) and sad where long>short ($p=.006$). We also found significant differences in happy where pat>contact ($p=.030$) and sad where contact>pat ($p=.001$).

12.4.3 Summary

Our experiment results showed that the effectiveness of the touch length and the touch type did not indicate the effectiveness of the touched part. A short touch was appropriate for happy emotions; a long touch was appropriate for sad emotions. Therefore, prediction 1 was supported. A pat was an appropriate touch type for happy emotions, and that contact was appropriate for sad emotions. Therefore, prediction 2 was supported. Finger touches are stronger and more natural than hand touches, regardless of the emotions. Therefore, prediction 3 was not supported.

12.5 DISCUSSION

Our experiment results identified the effectiveness of two touch characteristics (length and type) for expressing different emotions (happy and sad) for social robots. They show the advantages of finger touches over hand touches, results which are not necessarily consistent with past findings [13]. Possible reasons include the differences of robot appearance, touch feelings, and culture. For example, if we were to use robots with more robotic appearances, such as Pepper and Robovie, the effects of the characteristics would change. We only used a specific android with a feminine-like appearance; if we used an android robot with a masculine-like appearance [31,32] or a robot-like appearance [33–35], the perceived impressions would undoubtedly change. Investigating the effects of robot appearance among different countries is another intriguing future possibility.

In this study, we focused on a situation where a robot touches people, and our results identified several common characteristics with studies that focused on a situation where people touch a robot. One common characteristic is the touch length. A previous study reported that a relatively long touch is related to negative valence perspective [27], a finding confirmed by our study. Investigating the similarities and differences between touchers and receivers is another interesting future direction.

A further application is adapting different touch styles, such as hugging [36–38]. Because a past study focused on intra-hug gestures such as squeezing and rubbing, their touch characteristics seem related to conveyed emotions even through hugging interaction. Many hug interaction studies focused on the effects of behavior changes or mechanisms for natural hug interaction between robots and people. Of course, understanding touch characteristics within hug behavior will improve natural and affective interactions.

12.6 CONCLUSION

Since social robots need to appropriately convey their emotions to people for more natural and acceptable interaction, we focused on the effects of touch characteristics because touch interaction is essential in emotional contexts. We focused on conveying two typical emotions in human–robot interaction: happy and sad. We also focused on three different touch characteristics: length, type, and body part. We implemented different touch behaviors based on these emotions and characteristics in an android robot ERICA, which has a feminine-like appearance.

We experimentally investigated with human participants the relationships between perceived impressions of emotions expressed by the robot and its touch characteristics. Our experiment results showed that short and pat-type touches by fingers effectively convey strong and natural happy emotions. A long touch and a contact-type touch by fingers effectively conveys sad emotions.

ACKNOWLEDGMENT

This work was supported by JST CREST Grant Number JPMJCR18A1, Japan.

REFERENCES

[1] H. Felzmann, T. Beyan, M. Ryan, and O. Beyan, "Implementing an ethical approach to big data analytics in assistive robotics for elderly with dementia," ACS *SIGCAS Computers and Society*, vol. 45, no. 3, pp. 280–286, 2016.

[2] M. Shiomi, T. Iio, K. Kamei, C. Sharma, and N. Hagita, "Effectiveness of social behaviors for autonomous wheelchair robot to support elderly people in Japan," *PLoS One*, vol. 10, no. 5, p. e0128031, 2015.

[3] I. Leite, M. McCoy, M. Lohani, D. Ullman, N. Salomons, C. Stokes, S. Rivers, and B. Scassellati, "Narratives with robots: The impact of interaction context and individual Differences on story recall and emotional understanding," *Frontiers in Robotics and AI*, vol. 4, p. 29, 2017.

[4] M. Shiomi, T. Kanda, I. Howley, K. Hayashi, and N. Hagita, "Can a social robot stimulate science curiosity in classrooms?," *International Journal of Social Robotics*, vol. 7, no. 5, pp. 641–652, 2015.

[5] S. Satake, K. Hayashi, K. Nakatani, and T. Kanda, "Field trial of an information-providing robot in a shopping mall," In *The IEEE/RSJ International Conference on Intelligent Robots and Systems (IROS 2015)*, pp. 1832–1839, 2015.

[6] M. Shiomi, T. Kanda, D. F. Glas, S. Satake, H. Ishiguro, and N. Hagita, "Field trial of networked social robots in a shopping mall," In *2009 IEEE/RSJ International Conference on Intelligent Robots and Systems*, pp. 2846–2853, 2009.

[7] J. B. F. van Erp, and A. Toet, "Social touch in human-computer interaction," *Frontiers in Digital Humanities*, vol. 2, no. 2, p. 2, 2015.

[8] G. Huisman, "Social touch technology: A survey of haptic technology for social touch," *IEEE Transactions on Haptics*, vol. 10, no. 3, pp. 391–408, 2017.

[9] M. Shiomi, K. Shatani, T. Minato, and H. Ishiguro, "How should a robot react before people's touch?: Modeling a pre-touch reaction distance for a robot's face," *IEEE Robotics and Automation Letters*, vol. 3, no. 4, pp. 3773–3780, 2018.

[10] A. E. Block, and K. J. Kuchenbecker, "Softness, warmth, and responsiveness improve robot hugs," *International Journal of Social Robotics*, vol. 11, no. 1, pp. 49–64, 2019.

[11] J. Li, W. Ju, and B. Reeves, "Touching a mechanical body: Tactile contact with intimate parts of a humanoid robot is physiologically arousing," vol. 6, no. 3, pp. 118–130, 2017.

[12] Y. Yamashita, H. Ishihara, T. Ikeda, and M. Asada, "Appearance of a robot influences causal relationship between touch sensation and the personality impression," In *Proceedings of the 5th International Conference on Human Agent Interaction,* pp. 457–461, 2017.

[13] S. Yohanan, and K. E. MacLean, "The role of affective touch in human-robot interaction: Human intent and expectations in touching the haptic creature," *International Journal of Social Robotics*, vol. 4, no. 2, pp. 163–180, 2012.

[14] R. Lowe, R. Andreasson, B. Alenljung, A. Lund, and E. Billing, "Designing for a wearable affective interface for the NAO Robot: A study of emotion conveyance by touch," *Multimodal Technologies and Interaction*, vol. 2, no. 1, p. 2, 2018.

[15] B. Alenljung, R. Andreasson, R. Lowe, E. Billing, and J. Lindblom, "Conveying emotions by touch to the nao robot: A user experience perspective," *Multimodal Technologies and Interaction*, vol. 2, no. 4, p. 82, 2018.

[16] X. Meng, N. Yoshida, X. Wan, and T. Yonezawa, "Emotional gripping expression of a robotic hand as physical contact," In *Proceedings of the 7th International Conference on Human-Agent Interaction,* pp. 37–42, 2019.

[17] A. Ree, L. M. Mayo, S. Leknes, and U. Sailer, "Touch targeting C-tactile afferent fibers has a unique physiological pattern: A combined electrodermal and facial electromyography study," *Biological Psychology*, vol. 140, pp. 55–63, 2019.

[18] L. M. Mayo, J. Lindé, H. Olausson, and M. Heilig, "Putting a good face on touch: Facial expression reflects the affective valence of caress-like touch across modalities," *Biological Psychology*, vol. 137, pp. 83–90, 2018.

[19] D. F. Glas, T. Minato, C. T. Ishi, T. Kawahara, and H. Ishiguro, "Erica: The erato intelligent conversational android," In *2016 25th IEEE International Symposium on Robot and Human Interactive Communication (RO-MAN),* pp. 22–29, 2016.

[20] X. Zheng, M. Shiomi, T. Minato, and H. Ishiguro, "What kinds of robot's touch will match expressed emotions?," *IEEE Robotics and Automation Letters*, pp. 127–134, 2019.

[21] S. Rossi, F. Ferland, and A. Tapus, "User profiling and behavioral adaptation for HRI: A survey," *Pattern Recognition Letters*, vol. 99, pp. 3–12, 2017.

[22] I. Leite, C. Martinho, and A. Paiva, "Social robots for long-term interaction: A survey," *International Journal of Social Robotics*, vol. 5, no. 2, pp. 291–308, 2013.

[23] J. A. Russell, "A circumplex model of affect," *Journal of Personality and Social Psychology*, vol. 39, no. 6, p. 1161, 1980.

[24] M. Teyssier, G. Bailly, C. Pelachaud, and E. Lecolinet, "Conveying emotions through device-initiated touch," *IEEE Transactions on Affective Computing*, vol. 14, pp. 1477–1488, 2020.

[25] J. T. Suvilehto, E. Glerean, R. I. M. Dunbar, R. Hari, and L. Nummenmaa, "Topography of social touching depends on emotional bonds between humans," *Proceedings of the National Academy of Sciences* vol. 112, no. 45, pp. 13811–13816, 2015.

[26] T. Hirano, M. Shiomi, T. Iio, M. Kimoto, I. Tanev, K. Shimohara, and N. Hagita, "How do communication cues change impressions of human-robot touch interaction?," *International Journal of Social Robotics*, vol. 10, no. 1, pp. 21–31, 2018.

[27] T. W. Bickmore, R. Fernando, L. Ring, and D. Schulman, "Empathic touch by relational agents," *IEEE Transactions on Affective Computing*, vol. 1, no. 1, pp. 60–71, 2010.

[28] T. Fong, I. Nourbakhsh, and K. Dautenhahn, "A survey of socially interactive robots," *Robotics and Autonomous Systems*, vol. 42, no. 3–4, pp. 143–166, 2003.

[29] T. Kanda, R. Sato, N. Saiwaki, and H. Ishiguro, "A two-month field trial in an elementary school for long-term human-robot interaction," *IEEE Transactions on Robotics*, vol. 23, no. 5, pp. 962–971, 2007.

[30] T. Kanda, M. Shiomi, Z. Miyashita, H. Ishiguro, and N. Hagita, "A communication robot in a shopping mall," *IEEE Transactions on Robotics*, vol. 26, no. 5, pp. 897–913, 2010.

[31] D. Sakamoto, and H. Ishiguro, "Geminoid: Remote-controlled android system for studying human presence," *Kansei Engineering International*, vol. 8, no. 1, pp. 3–9, 2009.

[32] M. Shiomi, H. Sumioka, K. Sakai, T. Funayama, and T. Minato, "SŌTO: An android platform with a masculine appearance for social touch interaction," In *Companion of the 2020 ACM/IEEE International Conference on Human-Robot Interaction*, pp. 447–449, 2020.

[33] T. Kanda, H. Ishiguro, T. Ono, M. Imai, and R. Nakatsu, "Development and evaluation of an interactive humanoid robot "Robovie"," In *Proceedings 2002 IEEE International Conference on Robotics and Automation (Cat. No.02CH37292)*, vol. 2, pp. 1848–1855, 2002.

[34] R. Matsumura, M. Shiomi, K. Nakagawa, K. Shinozawa, and T. Miyashita, "A desktop-sized communication robot: "Robovie-mr2"," *Journal of Robotics and Mechatronics*, vol. 28, no. 1, pp. 107–108, 2016.

[35] R. Matsumura, and M. Shiomi, "An animation character robot that increases sales," *Applied Sciences*, vol. 12, no. 3, p. 1724, 2022.

[36] Y. Onishi, H. Sumioka, and M. Shiomi, "Increasing torso contact: Comparing human-human relationships and situations," In *International Conference on Social Robotics*, pp. 616–625, 2021.

[37] M. Shiomi, A. Nakata, M. Kanbara, and N. Hagita, "Robot reciprocation of hugs increases both interacting times and self-disclosures," *International Journal of Social Robotics*, pp. 1–9, 2020.

[38] M. Shiomi, and N. Hagita, "Audio-visual stimuli change not only robot's hug impressions but also its stress-buffering effects," *International Journal of Social Robotics*, vol. 13, pp. 469–476, 2021.

CHAPTER 13

Modeling Touch Timing and Length to Express Complex Emotions

Xiqian Zheng and Masahiro Shiomi

Advanced Telecommunications Research Institute International, Kyoto, Japan

Takashi Minato

RIKEN, Saitama, Japan

Hiroshi Ishiguro

Osaka University, Osaka, Japan

13.1 INTRODUCTION

In human–human interaction, touch behaviors have essential roles in the context of well-being, mental, and physical benefits [1–7]. Robotics researchers have also focused on touch behaviors for more natural and acceptable social robots in daily settings [8,9], as well as on their usefulness for conveying such intentions as emotions and intimacy to people by touch [10,11], because such capabilities can contribute to affective interaction with people. Based on these contexts, we want to understand what kinds of touch characteristics effectively convey emotions from robots to people.

Past studies investigating touch interaction characteristics focused on type, length, and place [10,11] to convey relatively simple emotions (happy/

DOI: 10.1201/9781003384274-17

sad) and show intimacy. Others focused on different characteristics and evaluated how touching behaviors changed the emotions conveyed by a robot [12–15]. Another study reported that touch interaction from a robot toward an object increased perceived kawaii feelings and touch from a human presenter [16]. These studies provided rich knowledge to convey robots' emotions to interacting people by touch interaction.

However, even though these studies identified the essential characteristics of emotional expressions by touch, the appropriate timing for conveying emotions and touch durations have received less attention. We believe that these characteristics are important to more naturally and strongly convey emotions. For example, negative emotions such as anger and fear elicit a relatively rapid reaction [17,18], which is related to proactive and short reactions, and therefore, people who are scared or surprised may touch others immediately and briefly. In addition, since the continuation time of heartwarming emotions is relatively long after their evocation compared to negative emotions [19,20], people who are moved may touch others after the timing of being moved. In the context of touch timing, a few studies investigated such effects, but they mainly focused on a robot's warmth or touching effects [21,22], rather than emotional naturalness and strength.

Therefore, in this study, we develop a model to control the appropriate timing to express robots' emotions by touch behaviors. For this purpose, we collected data about people's touch timing and durations when they express heartwarming and horror emotions using an android robot (Figure 13.1).

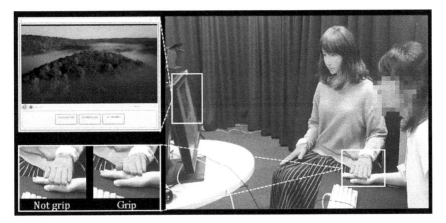

FIGURE 13.1 An experimental scene where ERICA and a participant watch a video [23].

Using these gathered data, we developed models that determine the timing and length of touch behaviors to express these two emotions by a robot's touch. We implemented the models in an android robot and confirmed that they enable the robot to select appropriate timing and durations in touch interaction. Note that this chapter is modified based on our previous work [23], edited to be comprehensive and fit with the context of this book.

13.2 TARGET EMOTIONS, ROBOT AND TOUCH BEHAVIOR

13.2.1 Target Emotion

Similar to a past study [11], this study focused on two different positive and negative emotions. First, we focused on heartwarming as a positive emotion because a past study concluded that Japanese people experience both happiness and sadness when they are moved [19]. As a counterpart to these positive emotions, we focused on horror as a negative emotion that combines fear and surprise.

To express such complex emotions by robots, people need to share their contexts with robots, such as observing the same visual stimuli. Moreover, thrilling emotions are related to the timing of specific situations, i.e., climax scenes. Based on these considerations, we decided on a situation where a robot and participants watched videos together because such situations enable the latter to easily imagine appropriate touch characteristics for robots that express emotions that appropriately match the visual stimuli.

13.2.2 Robot Setup

Robots need various capabilities to express emotions in interaction with people, such as facial expressions, speech synthesis, gestures, and touch behaviors. Therefore, we used an android robot, ERICA [24]. It has ten degrees of freedom in each arm and can control each actuator with 50-ms frequency, enabling it to express emotions by touch behaviors [10,11]. Even though the appearance of the silicone skin on ERICA's arms resembles human skin, its touch feels quite different, and therefore, we put gloves on ERICA when it touched the participants.

13.2.3 Touch Behavior Design

As its touch behavior, we employed a grip behavior with which the robot expressed its emotions by touching a participant's hand. This design followed past studies that investigated the touch effects of perceived emotions using robots [21,22,25] and a remote touch device [26]. These studies designed the robot's hand so that it is always touching a participant's hand.

Using grip behaviors provides simple control of the start/end timing of the touch behaviors. Related to this touch design, we also employed a situation where participants watched video stimuli with the robot, because in such a situation, a robot's expressed emotions related to the video contents will be naturally perceived by the participants. We placed the robot next to the participants. We also set markers on a table to fix the participants' hands during the experiment to reproduce similar touch behaviors. In summary, we collected data to find the appropriate grip timing and duration to express emotions while the robot watched video stimuli with the participants.

13.3 MATERIALS AND METHODS OF EXPERIMENT

13.3.1 Data Collection

13.3.1.1 Overview

In the data collection, we focused on three characteristics of gripping behaviors: the climax timing of the video watching, t_{climax}; the timing to start a grip behavior as a reaction (or anticipation) to the video's climax, t_{touch}; and the grip's duration, Δt. Our participants directly determined these three parameters to design the robot's touch behaviors. Based on part of this information, we calculated t_{start} (i.e., $t_{climax} - t_{touch}$), which is the difference between the touch and climax times (Figure 13.2).

The participants sequentially watched the video clips with the robot as many times as they wanted as they adjusted these characteristics to match their own preferences. We prepared three video clips of movie trailers or advertisements from YouTube[1] that reflected heartwarming and

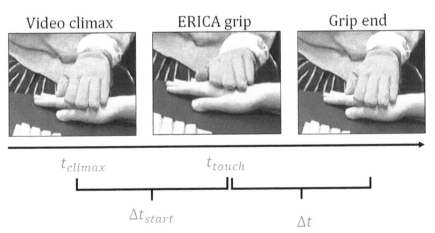

FIGURE 13.2   Illustration of t_{climax}, t_{touch}, Δt, and t_{start} [23]: (a) Heartwarming videos and (b) horror videos.

horror emotions. We edited their lengths between 98 and 159 s ($M = 118.3$ s, $SD = 26.2$ s) for the data collection.

13.3.1.2 Procedure

Before the experiment, we briefly interacted with the participants and explained the aims of the data collection: gathering timing data to express emotions by touch behaviors with a robot. The ethics committee of our institution approved this research.

Before the experiment started, the participants sat on the left of our robot, and we calibrated their hand position to decide the tables and markers for identical touch behaviors for all of the participants. In front of them, we placed a monitor that played the video stimuli and a user interface that controlled the robot's behaviors.

The participants can change the parameters of the robot's touch behavior by the user interface, i.e., t_{climax}, t_{touch}, and Δt. The robot records these input values by the participants for the data collection. Once the participants satisfied the adjusted parameters to express emotions via touch behaviors of the robot, they repeated the procedure for the remaining clips. We adopted a counterbalance design to play either the first three horror videos or the heartwarming videos and then vice versa.

13.3.1.3 Participants

Forty-eight participants (24 women and 24 men) joined this study. Their ages ranged from 20 to 49. We asked the participants whether they had watched the movies used in the experiment beforehand and verified that they had never experienced any touch interactions with a robot.

13.3.2 Hypotheses and Predictions

This study investigated the differences between touch characteristics (timing and durations) for expressing positive and negative emotions. Past studies reported that the continuation time of heartwarming emotion, a positive emotion, is longer than negative emotions [19,20]. On the other hand, other studies reported that people rapidly respond to negative emotional stimulus [17,18]. Therefore, we hypothesized that the grip-behavior timing for heartwarming emotions will start later than the grip-behavior timing for horror emotions. In addition, the grip-behavior duration for heartwarming emotions will be longer than for horror emotions. Based on these hypotheses, we made the following two predictions:

Prediction 1: The t_{start} of the heartwarming emotions will be later than the t_{start} for horror emotions.

Prediction 2: The Δt for heartwarming emotions will be longer than the Δt for horror emotions.

13.4 RESULTS

We gathered 288 t_{climax}, t_{touch}, and Δt items and calculated the t_{start} data from the data collection. 20 items were excluded due to hardware problems. We also used the Z scores of each item to find outliers. After excluding two more items, our modeling process contained 262 items.

13.4.1 Analysis of t_{climax}, t_{touch}, t_{start}, and Δt

Although a few videos have different climax timings, most of the participants share similar t_{climax} timings. In addition, participants selected slightly shifted later t_{touch} timings for the heartwarming videos and earlier for the horror timing compared to t_{climax}. Even though a part of the participants found different climax timings in the videos, their timing design was similar.

Figures 13.3 and 13.4 show the t_{start} and Δt histograms for all the heartwarming/horror videos. Participants assigned different grip timings due

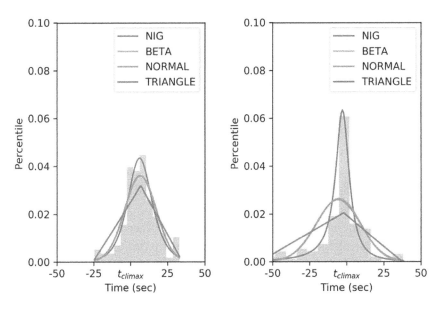

FIGURE 13.3   Histogram and fitting results of t_{start} [23]: (a) Heartwarming videos and (b) horror videos.

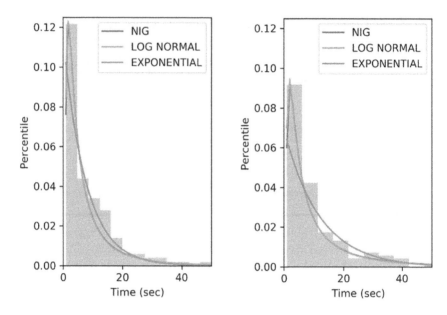

FIGURE 13.4   Histogram and fitting results of Δt [23].

to the video types. The start timing's peak grip behavior for the heart-warming videos is relatively later, and the peak for the horror videos is relatively earlier.

13.4.2 Statistical Analysis

We conducted a one-way repeated ANOVA to investigate the differences in the start timing and the duration of the touch behaviors between the positive/negative emotions. The results for t_{start} showed a significant difference in the video category factor (F (1, 132) = 33.797, $p<0.001$, partial $\eta^2 = 0.204$). We also conducted a one-way repeated measures ANOVA for Δt, whose results showed a significant difference in the video category factor (F (1, 132) = 7.226, $p=0.008$, partial $\eta^2 = 0.052$). Thus, t_{start} and Δt are significantly different between the heartwarming and horror videos. Thus, predictions 1 and 2 are supported.

As an additional analysis, we investigated whether the grip behaviors continued during the climax timing when the grip timing came before the climax. The number of cases for heartwarming videos is 20/27, whereas the number of cases for horror videos is 56/76. The binominal test showed that most touch durations lasted beyond the climax timing (heartwarming: $p = 0.019$, horror: $p < 0.001$), suggesting that maintaining a grip beyond the climax timing is useful for reproducing human-like touch behavior.

13.4.3 Modeling Grip Timing

Based on the gathered data from the data collection, we employed a fitting approach with probabilistic functions for mathematical modeling of the gripping behaviors. We first compared the t_{start} histograms and the probability distribution functions, including normal, beta, triangle, and normal-inversed Gaussian (NIG). Then, we calculated their R-squared values (R^2). We chose NIG with the parameters because it showed a higher R^2 than the other functions.

13.4.4 Modeling Touch Duration

Our experiment results showed that the touch durations are different for heartwarming and horror emotions. The participants set relatively longer durations for the heartwarming emotions and shorter durations for the horror emotions. In the modeling process of these touch durations, we investigated which probability distribution functions fit the data distribution. Since NIG is also suitable for model fitting, we again employed the NIG model to calculate the touch durations.

13.4.5 Implementation

Finally, we used the modeled grip timing and duration to implement appropriate touch behaviors with ERICA. It can decide t_{start} based on the NIG model due to the emotion category (heartwarming or horror) and the pre-defined t_{climax} of the target video. It can also decide the duration of its own grip timing (Δt) based on the NIG model. We confirmed that the robot can autonomously decide its grip timing and durations based on the video information.

13.5 DISCUSSION

This study investigated appropriate touch parameters for conveying complex emotions to people by gripping behaviors. One possible application is investigating such parameters for different touch interactions, such as hugging [27–29]. Although these past studies showed the effectiveness of hug interaction, they focused less on detailed parameter modeling to convey emotions by hugging. Moreover, cultural differences may have influences toward touch interaction [30,31], as well as touch contexts [32,33] and emotional expressions in non-verbal behaviors [34,35].

This study has several limitations since it used a specific robot (ERICA) and a specific touch (gripping). If we conducted data collection with different robots [36–38] or androids with quite human-like appearances [24,39], we would undoubtedly identify different parameters.

13.6 CONCLUSION

We analyzed the appropriate grip timing and duration for conveying two complex emotions (heartwarming and horror) by conducting a data collection with human participants. From the gathered data, 48 participants showed different grip timings and durations based on the emotional category. We used NIG distribution functions to model appropriate grip timing and durations for both heartwarming and horror emotions. We implemented these models to autonomously convey different emotions through grip behavior toward a human-like appearance robot, and we confirmed that our developed system appropriately determined the characteristics of grip behaviors depending on the categories of visual stimuli.

ACKNOWLEDGMENT

This work was supported by JST CREST Grant Number JPMJCR18A1, Japan.

NOTE

1 https://youtu.be/PFhQhpR5Z8M, https://youtu.be/r1gz-m5Ai_E, https://youtu.be/b2MH-yxIR4Y, https://youtu.be/4LYK0rTjlM8, https://youtu.be/dCPiAOiKSyo, https://youtu.be/gXfLl3qYy0k.

REFERENCES

[1] A. Cekaite, and M. Kvist Holm, "The comforting touch: Tactile intimacy and talk in managing children's distress," *Research on Language and Social Interaction*, vol. 50, no. 2, pp. 109–127, 2017.

[2] C. O'lynn, and L. Krautscheid, "'How should I touch you?': A qualitative study of attitudes on intimate touch in nursing care," *The American Journal of Nursing*, vol. 111, no. 3, pp. 24–31, 2011.

[3] B. K. Jakubiak, and B. C. Feeney, "Interpersonal touch as a resource to facilitate positive personal and relational outcomes during stress discussions," *Journal of Social and Personal Relationships*, vol. 36, no. 9, pp. 2918–2936, 2019.

[4] S. Cohen, D. Janicki-Deverts, R. B. Turner, and W. J. Doyle, "Does hugging provide stress-buffering social support? A study of susceptibility to upper respiratory infection and illness," *Psychological Science*, vol. 26, no. 2, pp. 135–147, 2015.

[5] B. K. Jakubiak, and B. C. Feeney, "Keep in touch: The effects of imagined touch support on stress and exploration," *Journal of Experimental Social Psychology*, vol. 65, pp. 59–67, 2016.

[6] A. Gallace, and C. Spence, "The science of interpersonal touch: An overview," *Neuroscience & Biobehavioral Reviews*, vol. 34, no. 2, pp. 246–259, 2010.

[7] K. C. Light, K. M. Grewen, and J. A. Amico, "More frequent partner hugs and higher oxytocin levels are linked to lower blood pressure and heart rate in premenopausal women," *Biological psychology*, vol. 69, no. 1, pp. 5–21, 2005.

[8] J. B. F. van Erp, and A. Toet, "Social touch in human-computer interaction," *Frontiers in Digital Humanities*, vol. 2, no. 2, p. 2, 2015.

[9] M. Shiomi, H. Sumioka, and H. Ishiguro, "Survey of social touch interaction between humans and robots," *Journal of Robotics and Mechatronics*, vol. 32, no. 1, pp. 128–135, 2020.

[10] X. Zheng, M. Shiomi, T. Minato, and H. Ishiguro, "How can robot make people feel intimacy through touch?," *Journal of Robotics and Mechatronics*, vol. 32, no. 1, pp. (to appeear), 2019.

[11] X. Zheng, M. Shiomi, T. Minato, and H. Ishiguro, "What kinds of robot's touch will match expressed emotions?," *IEEE Robotics and Automation Letters*, vol. 5, pp. 127–134, 2019.

[12] B. Alenljung, R. Andreasson, R. Lowe, E. Billing, and J. Lindblom, "Conveying emotions by touch to the nao robot: A user experience perspective," *Multimodal Technologies and Interaction*, vol. 2, no. 4, p. 82, 2018.

[13] R. Lowe, R. Andreasson, B. Alenljung, A. Lund, and E. Billing, "Designing for a wearable affective interface for the NAO Robot: A study of emotion conveyance by touch," *Multimodal Technologies and Interaction*, vol. 2, no. 1, p. 2, 2018.

[14] R. Andreasson, B. Alenljung, E. Billing, and R. Lowe, "Affective touch in human-robot interaction: Conveying emotion to the nao robot," *International Journal of Social Robotics*, vol. 10, no. 4, pp. 473–491, 2018.

[15] M. Teyssier, G. Bailly, C. Pelachaud, and E. Lecolinet, "Conveying emotions through device-initiated touch," *IEEE Transactions on Affective Computing*, vol. 13, pp. 1477–1488, 2020.

[16] Y. Okada, M. Kimoto, T. Iio, K. Shimohara, H. Nittono, and M. Shiomi, "Kawaii emotions in presentations: Viewing a physical touch affects perception of affiliative feelings of others toward an object," *PLoS One*, vol. 17, no. 3, p. e0264736, 2022.

[17] M. Mather, and M. R. Knight, "Angry faces get noticed quickly: Threat detection is not impaired among older adults," *The Journals of Gerontology Series B: Psychological Sciences and Social Sciences*, vol. 61, no. 1, pp. P54–P57, 2006.

[18] E. Fox, L. Griggs, and E. Mouchlianitis, "The detection of fear-relevant stimuli: Are guns noticed as quickly as snakes?," *Emotion*, vol. 7, no. 4, p. 691, 2007.

[19] A. Tokaji, "Research for determinant factors and features of emotional responses of "kandoh" (the state of being emotionally moved)," *Japanese Psychological Research*, vol. 45, no. 4, pp. 235–249, 2003.

[20] T. Takada, and S. Yuwaka, "Persistence of emotions experimentally elicited by watching films," *Bulletin of Tokai Gakuen University*, vol. 25, 2020.

[21] C. J. A. M. Willemse, D. K. J. Heylen, and J. B. F. van Erp, "Communication via warm haptic interfaces does not increase social warmth," *Journal on Multimodal User Interfaces*, vol. 12, no. 4, pp. 329–344, 2018.

[22] C. J. A. M. Willemse, A. Toet, and J. B. F. van Erp, "Affective and behavioral responses to robot-initiated social touch: Toward understanding the opportunities and limitations of physical contact in human-robot interaction," *Frontiers in ICT*, vol. 4, p. 12, 2017.

[23] X. Zheng, M. Shiomi, T. Minato, and H. Ishiguro, "Modeling the timing and duration of grip behavior to express emotions for a social robot," *IEEE Robotics and Automation Letters*, vol. 6, no. 1, pp. 159–166, 2020.

[24] D. F. Glas, T. Minato, C. T. Ishi, T. Kawahara, and H. Ishiguro, "Erica: The erato intelligent conversational android," In *2016 25th IEEE International Symposium on Robot and Human Interactive Communication (RO-MAN)*, New York, NY, United States, pp. 22–29, 2016.

[25] H. Kawamichi, R. Kitada, K. Yoshihara, H. K. Takahashi, and N. Sadato, "Interpersonal touch suppresses visual processing of aversive stimuli," *Frontiers in Human Neuroscience*, vol. 9, p. 164, 2015.

[26] J. Cabibihan, and S. S. Chauhan, "Physiological responses to affective tele-touch during induced emotional stimuli," *IEEE Transactions on Affective Computing*, vol. 8, no. 1, pp. 108–118, 2017.

[27] Y. Onishi, H. Sumioka, and M. Shiomi, "Increasing torso contact: Comparing human-human relationships and situations," In *International Conference on Social Robotics*, Singapore, Singapore, pp. 616–625, 2021.

[28] M. Shiomi, A. Nakata, M. Kanbara, and N. Hagita, "Robot reciprocation of hugs increases both interacting times and self-disclosures," *International Journal of Social Robotics*, vol. 13, pp. 353–361, 2020.

[29] M. Shiomi, and N. Hagita, "Audio-visual stimuli change not only robot's hug impressions but also its stress-buffering effects," *International Journal of Social Robotics*, vol. 13, pp. 469–476, 2021.

[30] E. McDaniel, and P. A. Andersen, "International patterns of interpersonal tactile communication: A field study," *Journal of Nonverbal Behavior*, vol. 22, no. 1, pp. 59–75, 1998.

[31] R. Dibiase, and J. Gunnoe, "Gender and culture differences in touching behavior," *The Journal of Social Psychology*, vol. 144, no. 1, pp. 49–62, 2004.

[32] J. A. Hall, "Touch, status, and gender at professional meetings," *Journal of Nonverbal Behavior*, vol. 20, no. 1, pp. 23–44, 1996.

[33] F. N. Willis, and L. F. Briggs, "Relationship and touch in public settings," *Journal of Nonverbal Behavior*, vol. 16, no. 1, pp. 55–63, 1992.

[34] P. Ekman, W. V. Friesen, M. O'sullivan, A. Chan, I. Diacoyanni-Tarlatzis, K. Heider, R. Krause, W. A. LeCompte, T. Pitcairn, and P. E. Ricci-Bitti, "Universals and cultural differences in the judgments of facial expressions of emotion," *Journal of Personality and Social Psychology*, vol. 53, no. 4, p. 712, 1987.

[35] D. Matsumoto, and H. S. C. Hwang, "Culture, emotion, and expression," In *Cross-Cultural Psychology: Contemporary Themes and Perspectives*, Wiley Blackwell, pp. 501–515, 2019.

[36] T. Kanda, H. Ishiguro, T. Ono, M. Imai, and R. Nakatsu, "Development and evaluation of an interactive humanoid robot "Robovie"," In *Proceedings 2002 IEEE International Conference on Robotics and Automation (Cat. No.02CH37292)*, Washington, DC, United States, vol. 2, pp. 1848–1855, 2002.

[37] R. Matsumura, M. Shiomi, K. Nakagawa, K. Shinozawa, and T. Miyashita, "A desktop-sized communication robot: "Robovie-mr2"," *Journal of Robotics and Mechatronics*, vol. 28, no. 1, pp. 107–108, 2016.

[38] R. Matsumura, and M. Shiomi, "An animation character robot that increases sales," *Applied Sciences*, vol. 12, no. 3, p. 1724, 2022.

[39] M. Shiomi, H. Sumioka, K. Sakai, T. Funayama, and T. Minato, "SŌTO: An android platform with a masculine appearance for social touch interaction," In *Companion of the 2020 ACM/IEEE International Conference on Human-Robot Interaction*, Cambridge, United Kingdom, pp. 447–449, 2020.

SECTION 5

Behavior Change Effects in Human–Robot Touch Interaction

CHAPTER 14

Robot Hugs Encourage Self-disclosures

Masahiro Shiomi

Advanced Telecommunications Research Institute International, Kyoto, Japan

Aya Nakata

Nara Institute of Science and Technology, Nara, Japan

Masayuki Kanbara

Nara Institute of Science and Technology, Nara, Japan

Norihiro Hagita

Osaka University of Arts, Osaka, Japan

14.1 INTRODUCTION

In the human–robot interaction research field, social bonding between robots and people is crucial for smooth interaction in the context of education [1–5], elderly care [6–8], hospitals [9,10], and shopping [11,12]. For this purpose, robotics researchers have focused on interaction strategies that achieve acceptable social robots by considering behavioral changes during interaction [12,13], positioning behaviors [14,15], and their characteristics [16–18].

Self-disclosure plays an essential role in social bonding [19,20]. In human science literature, researchers have identified the effectiveness of physical

DOI: 10.1201/9781003384274-19

FIGURE 14.1   Hug interaction with a robot.

interaction on eliciting self-disclosures [19,21]. Past studies focused on the interaction design of interactive agents to elicit self-disclosures from people, not physical interaction; instead, they focused on conversational interaction [19,22,23]. Some studies investigated the effectiveness of human–robot touch interaction in the context of behavior changes [24,25] and emotional expressions [26–29], although these studies focused less on the perspective of social bonding.

If social robots can efficiently use their own physical body for physical interaction and elicit self-disclosures, they will have an advantage in building social relationships with interacting people. Based on these considerations, we experimentally investigated the relationships between physical interactions with robots and people's self-disclosures with a robot that can hug people (Figure 14.1). Note that this chapter is modified based on our previous work [30], edited to be comprehensive and fit with the context of this book.

14.2 SYSTEM

14.2.1 Robot Hardware and Software

We used Moffuly-I, which we previously developed [31]. This robot is 200-cm tall and has two 80-cm arms (one degree of freedom for each elbow) for giving a hug. Each arm has a weak digital servo motor to move the elbow joints and fabric-based touch sensors for safety. Its frame is covered with polypropylene and a fluffy-material-based skin like in commercial dolls. We controlled the robot with a Wizard-of-Oz approach. An operator observed the robot and the interacting people by a camera and microphone to decide its behaviors.

14.2.2 Conversational Behavior

In this study, we examined the impact of hugging interactions on self-disclosure promotion using three distinct conversational behaviors: the robot's self-introduction, soliciting self-disclosures, and filler responses. First, the robot introduces itself, shares a personal detail, and moves its arm: "Hello, I'm Moffuly. Although I resemble a bear, my preferred sustenance is electricity, not honey." The robot encourages self-disclosure by listening attentively to the participants. Filler responses bridge gaps in the conversation. The specifics of the hugging behaviors will be elaborated below.

In our investigation, an operator selectively and partially teleoperated the robot in accordance with Wizard-of-Oz methodology [32] to select suitable conversational behaviors consisting of relatively straightforward utterances. The robot assumed the role of an empathetic listener to inspire self-disclosures from the participants. Consequently, if participants posed excessive inquiries of the robot, it might meekly protest: "Sorry, but that question is too challenging for me. I am more interested in learning about you." This tactic prevents the conversations from being centered on the robot. The objective of these rule implementations is to mitigate the influence of varying interaction styles between the participants and the robot. For teleoperation purposes, we positioned two cameras in the experimental space: one equipped with a microphone near the robot and another overhead. The operator eavesdropped on the participants with a microphone. To facilitate analysis, we documented video/audio data from the cameras and the microphone.

14.2.3 Hug Behavior

To execute the embracing interactions with the robot, Moffuly initially seeks a hug from the participants within our study's parameters (Section 4.3). To begin the interactions that will encompass the hugs, the robot opens its arms and asks, "Before starting our dialogue, would you please give me with a hug?" Before the participant's embrace, the robot closes its arms until it senses contact between them and the person's body. Following the next guidelines, the robot gently pats the participant's back by maneuvering both arms at the end of its own discourse or that of the participant. If the participant's speech extends beyond 30 seconds, the robot pats at 30-second intervals.

14.3 EXPERIMENT

14.3.1 Hypothesis and Predictions

Physical interactions with others generate favorable impressions and stimulate increased self-disclosure in human-to-human communication [19,21]. Similarly, physical interactions with robots in human–robot interactions also yield positive impressions [25,33–36]. As a result, we posit that individuals embraced by a robot will reveal more about themselves than those who are not hugged. Past studies concluded that reciprocal touch interactions with a robot evoke stronger emotions and prompt more behavioral changes compared to individuals who do not experience such reciprocal touches from a robot [25,37]. Consequently, we suggest that individuals hugged by a robot will disclose more about themselves than those who only hug it.

We hypothesized that people's inclination to engage with the robot will be influenced by physical interaction. We surmised that the robot's reciprocal hugs are instrumental in forming strong bonds. If this conjecture is accurate, those who are hugged by a robot will have more prolonged interactions, leading to increased self-disclosures. Thus, we formulated two predictions:

Prediction 1: Reciprocal hugs from a robot will lengthen the participant interaction times compared to interactions without a reciprocal hug or any physical contact.

Prediction 2: Reciprocal hugs from a robot will prompt greater self-disclosures from participants than interactions without reciprocal hugs or any physical contact.

14.3.2 Participants

Forty-eight Japanese individuals (24 women and 24 men whose average age was 36.19 with a standard deviation (SD) of 9.93 and a range from 20 to 52) received compensation for their involvement.

14.3.3 Conditions

The study employed a between-participant design featuring three distinct conditions. We assigned 16 participants (eight women and eight men) to each condition. An operator controlled the robot, adhering to the same rules and maintaining identical conversational content across the three conditions:

No-hug: Participants remained in their initial position (45 cm from the robot) and engaged in conversation without any physical interaction.

Hug-only: The robot asked for a hug from the participants before initiating a conversation without reciprocating.

Reciprocated hug: The robot requested a hug from the participants, reciprocated, and began talking. The robot occasionally patted them on the back during the experiment, following the pre-defined rules.

14.3.4 Procedures

The experimenter briefly described the experiment's purpose and procedure prior to its start. For those in the hug-only and reciprocated hug conditions, the experimenter physically demonstrated how to hug the robot and also explained its limited conversational capabilities, mentioning that it prefers listening to stories and engaging in simple conversations. After the experimenter left the room, the experiment began.

Upon starting, the robot greeted the participants and, in the hug-only or reciprocated hug conditions, requested a hug. It introduced itself, asked for a self-disclosure, and invited the participants to share a story. The experiment had a minimum duration of ten minutes, starting after the robot's finished its self-introduction. After ten minutes, participants took a short break and decided whether to conclude the experiment or to extend it for a maximum of ten more minutes during which time they could stop the interaction at any moment.

Before the experiment, the participants were told that the robot was autonomous. Following it, a debriefing session clarified our research purpose. All the participants believed that the robot was autonomous, most likely due to its simple utterances and reactions during the interactions.

Our institution's ethics committee approved this research involving human participants, and informed consent was obtained from every individual.

14.3.5 Measurements

We assessed two objective factors and one subjective factor. For the former type, we examined the participants' engagement durations and compared the proportions of self-disclosure to non-self-disclosure dialogues by dividing the self-disclosure-related dialogues by those without self-disclosures. We concentrated on these proportions because the participants' interaction durations and the conversations' content lengths varied.

To quantify both types of conversational content, a coder transcribed all the conversations from the recorded video/audio data and divided the conversation data into 289 segments. We defined a conversation unit as a conversational topic. The coder transcribed and segmented the texts based on the changes in the conversational topics. Consequently, when a participant discussed topics A, B, C, and then returned to A, the number of segments totaled four. Each segment consisted of several sentences.

Subsequently, the coder classified all the segments as either self-disclosure or non-self-disclosure. If the segments contained personal topics like hobbies or experiences, they were categorized as self-disclosure. If they only discussed mundane topics like the weather, they were considered non-self-disclosure. Additionally, the coder coded the self-disclosure content into positive/negative categories to examine whether the robot's physical interaction influenced the self-disclosure types. Following this process, another coder coded 10% of the data. We calculated the coding validity based on a prior study [38], and a kappa coefficient [39] of 0.71 indicated substantial agreement between the coders.

For the subjective metric, we gauged the participants' perceived positive impressions of the robot using a single questionnaire item: "I think this robot is good overall." Participants rated this item on a 1-to-7 point scale, where 1 represented the most negative response (complete disagreement) and 7 was the most positive response (complete agreement).

14.4 RESULTS

14.4.1 Verification of Prediction 1

Figure 14.2 illustrates the interaction durations of the participants. To analyze the data, we carried out an ANOVA, which revealed significant differences ($F (2, 45) = 18.030$, $p < 0.001$, partial $\eta^2 = 0.445$). Multiple comparisons employing the Bonferroni method revealed significant differences among the conditions: *reciprocated hug > hug-only* ($p < 0.001$) and *reciprocated hug > no-hug* ($p < 0.001$). No significant difference was observed between *hug-only* and *no-hug* ($p = 1.000$). Consequently, the robot's reciprocated hugs significantly increased the interaction length more than without them and without any physical interaction. Thus, prediction 1 was supported.

14.4.2 Verification of Prediction 2

Figure 14.3 shows the proportions of self-disclosure to the non-self-disclosure dialogues. To analyze the data, we carried out an ANOVA, which revealed significant differences ($F (2, 45) = 8.162$, $p = 0.001$, partial $\eta^2 = 0.266$). Multiple comparisons employing the Bonferroni method

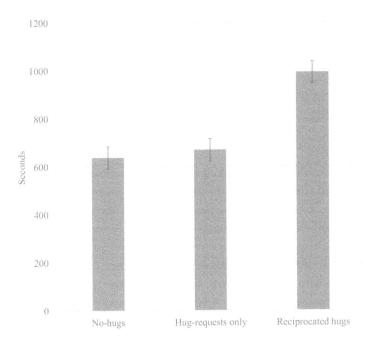

FIGURE 14.2   The interaction duration in each condition.

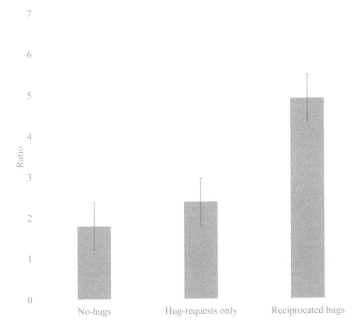

FIGURE 14.3   The proportion of self-disclosure to the non-self-disclosure dialogues in each condition.

revealed significant differences: *reciprocated hug>hug-only* ($p=0.010$), *reciprocated hug>no-hug* ($p=0.001$). No significant difference was observed between *hug-only* and *no-hug* ($p=1.000$). Consequently, the robot's reciprocated hugs significantly increased the ratio of self-disclosures more than without the reciprocated hugs and any physical interaction. Thus, prediction 2 was supported.

14.4.3 Analysis of Total Self-Disclosure Amount

Through our evaluation, the reciprocated hugs from the robot effectively enhanced the interaction duration and the proportions of both the self-disclosure and non-self-disclosure dialogues. We also compared the quantities of self-disclosure and non-self-disclosure conversations among the conditions.

Figure 14.4 shows the number of both the self-disclosure and non-self-disclosure conversations. To analyze the data, we conducted a two-way repeated measure ANOVA with mixed factors: category (self-disclosure and non-self-disclosure) and condition (*no-hug, hug-only,* and *reciprocated hug*). Our results showed significant differences in the category factor (F (1, 45)$=21.378$, $p<0.001$, partial $\eta^2=0.322$) and the

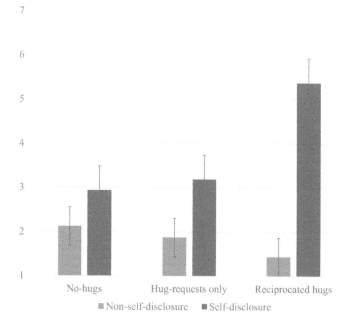

FIGURE 14.4 The number of both the self-disclosure and non-self-disclosure conversations in each condition.

interaction between the two factors (F (2, 45)=4.081, p=0.024, partial η^2=0.154). No significant variation was observed in the condition factor (F (2, 45)=2.297, p=0.112).

Multiple comparisons with the Bonferroni method showed significant differences in the *self-disclosure* category (*reciprocated hug>hug-only*, p=0.039) and in the *reciprocated hug* condition (*self-disclosure>non-self-disclosure*, p<0.001). The results showed significant trends in the *self-disclosure* category (*reciprocated hug>no-hug*, p=0.073) and in the *hug-only* condition (*self-disclosure>non-self-disclosure*, p=0.078). These results imply that the robot's reciprocated hugs significantly increased the amount of self-disclosures.

14.5 DISCUSSION

14.5.1 What Led the Reciprocated Hugs to Enhance the Interaction Duration and the Number of Self-Disclosures?

Our experimental findings indicate that individuals interacted with the robot for a longer period and offered increased self-disclosures when it provided reciprocal hugs. These experiment results prompt the following inquiry: Which occurred first, the cause or the effect? It is challenging to determine whether they interacted more due to engaging in self-disclosure or whether they disclosed more because they spent more time with the robot.

We posit that both factors contribute, although the influence of the reciprocated hugs on self-disclosure is more pronounced for a couple of reasons. First, we examined the ratios of the self-disclosure to the non-self-disclosure conversations that had a minimum length of ten minutes. The outcomes did not show significant differences between conditions, indicating that participants engaged in more self-disclosure during the prolonged interactions. Since the experiment demonstrated that interaction time significantly increased in the reciprocated hug condition, the hugs seemed to contribute to longer interaction durations and potentially promoted self-disclosure. Second, basic physical responses (e.g., patting during a hug) could be interpreted as supplementary social cues in conversations, fostering greater conversational engagement from participants. Naturally, this interactive loop stimulates self-disclosures and heightens the desire to interact. Additionally, previous research highlighted the persuasive effects of active touch interactions [40]. Thus, reciprocated hugs might amplify such effects when soliciting self-disclosures.

Our experimental findings revealed that participants engaged longer in the reciprocated hug condition than the other conditions. This result might be attributed to the variations in the responses of the participants since the robot did not ask for a hug in the no-hug condition. Despite that, the actions requested by the participants were consistent between the reciprocated hug and hug-only conditions, as was the robot's pre-experiment speech (e.g., "please give me a hug"). Thus, its reciprocal hug might have influenced the interaction duration.

14.5.2 Limitations

Our study has some limitations due to the specific robot used in our experiments, a situation that restricts the generality of our findings. To apply our insights to various scenarios, it is crucial to account for factors such as the robot's size and the nature of its touch. Moreover, our implementation featured a relatively simple hug interaction; of course, humans exhibit a wide range of hug types. Investigating the effects of different hug interactions is essential to deepening our understanding of the impact of robot hugs, as a previous study reported that touch characteristics can alter the perceived impressions in human–robot touch interactions [26].

Despite these limitations, the knowledge gained from our research has valuable implications for the field of human–robot touch interactions. One potential application lies in clinical settings, where reciprocated hug interactions could foster rapport between robots and patients. A previous study found that patients preferred disclosing personal information to computer graphics (CG)-based agents rather than teleoperated ones [41]. Our research highlights the effectiveness of physical interaction, an aspect that CG-based agents cannot replicate. Consequently, hug interactions with robots may contribute to building relationships with patients through self-disclosures.

14.6 CONCLUSION

Although prior research demonstrated the benefits of human-to-robot touch interactions, the impact of robot-to-human reciprocal touch remains less explored. Our study focused on the effects of a social robot's reciprocated hugs on promoting self-disclosures and fostering increased interaction. We conducted an experiment using a robot capable of hugging human participants. We found that participants who experienced a hug from the robot engaged in longer interactions than those who did not receive a hug. Moreover, the reciprocated hug group disclosed significantly more personal information than their counterparts in the non-hug condition. This evidence highlights the value of robot-initiated haptic interactions in human–robot relationship development.

Note, however, that the participants' positive impressions of the robot did not significantly differ between conditions, despite the observed behavioral differences. Factors such as minimum interaction time or the robot's appearance and voice may have contributed to these perceptions, although it remains an open question. Future research should explore the effects of haptic interactions on perceived impressions using various measurement methods, such as hormone levels or brain activity.

ACKNOWLEDGMENT

This work was supported by JST CREST Grant Number JPMJCR18A1, Japan.

REFERENCES

[1] S.-S. Yun, M. Kim, and M.-T. Choi, "Easy interface and control of tele-education robots," *International Journal of Social Robotics*, vol. 5, no. 3, pp. 335–343, 2013.

[2] M. Shiomi, T. Kanda, I. Howley, K. Hayashi, and N. Hagita, "Can a social robot stimulate science curiosity in classrooms?," *International Journal of Social Robotics*, vol. 7, no. 5, pp. 641–652, 2015.

[3] T. Komatsubara, M. Shiomi, T. Kanda, H. Ishiguro, and N. Hagita, "Can a social robot help children's understanding of science in classrooms?," In *Proceedings of the 2nd International Conference on Human-Agent Interaction*, Tsukuba, Japan, pp. 83–90, 2014.

[4] F. Tanaka, K. Isshiki, F. Takahashi, M. Uekusa, R. Sei, and K. Hayashi, "Pepper learns together with children: Development of an educational application," In *2015 IEEE-RAS 15th International Conference on Humanoid Robots (Humanoids)*, Seoul, Korea, pp. 270–275, 2015.

[5] F. Tanaka, and S. Matsuzoe, "Children teach a care-receiving robot to promote their learning: Field experiments in a classroom for vocabulary learning," *Journal of Human-Robot Interaction*, vol. 1, no. 1, 2012.

[6] H. Felzmann, T. Beyan, M. Ryan, and O. Beyan, "Implementing an ethical approach to big data analytics in assistive robotics for elderly with dementia," *ACM SIGCAS Computers and Society*, vol. 45, no. 3, pp. 280–286, 2016.

[7] E. Mordoch, A. Osterreicher, L. Guse, K. Roger, and G. Thompson, "Use of social commitment robots in the care of elderly people with dementia: A literature review," *Maturitas*, vol. 74, no. 1, pp. 14–20, 2013.

[8] M. Shiomi, T. Iio, K. Kamei, C. Sharma, and N. Hagita, "Effectiveness of social behaviors for autonomous wheelchair robot to support elderly people in Japan," *PLoS One*, vol. 10, no. 5, p. e0128031, 2015.

[9] B. Mutlu, and J. Forlizzi, "Robots in organizations: The role of workflow, social, and environmental factors in human-robot interaction," In *Proceedings of the 3rd ACM/IEEE International Conference on Human Robot Interaction*, Amsterdam, The Netherlands, pp. 287–294, 2008.

[10] S. Ljungblad, J. Kotrbova, M. Jacobsson, H. Cramer, and K. Niechwiadowicz, "Hospital robot at work: Something alien or an intelligent colleague?," In *Proceedings of the ACM 2012 conference on Computer Supported Cooperative Work*, New York, NY, United States, pp. 177–186, 2012.

[11] Y. Iwamura, M. Shiomi, T. Kanda, H. Ishiguro, and N. Hagita, "Do elderly people prefer a conversational humanoid as a shopping assistant partner in supermarkets?," In *2011 6th ACM/IEEE International Conference on Human-Robot Interaction (HRI)*, Lausanne, Switzerland, pp. 449–457, 2011.

[12] T. Kanda, M. Shiomi, Z. Miyashita, H. Ishiguro, and N. Hagita, "A communication robot in a shopping mall," *IEEE Transactions on Robotics*, vol. 26, no. 5, pp. 897–913, 2010.

[13] T. Kanda, R. Sato, N. Saiwaki, and H. Ishiguro, "A two-month field trial in an elementary school for long-term human-robot interaction," *IEEE Transactions on Robotics*, vol. 23, no. 5, pp. 962–971, 2007.

[14] J. Mumm, and B. Mutlu, "Human-robot proxemics: Physical and psychological distancing in human-robot interaction," In *Proceedings of the 6th International Conference on Human-Robot Interaction*, Lausanne, Switzerland, pp. 331–338, 2011.

[15] C.-M. Huang, T. Iio, S. Satake, and T. Kanda, "Modeling and controlling friendliness for an interactive museum robot," In *Robotics: Science and Systems*, Berkeley, California, United States, pp. 12–16, 2014.

[16] I. Kruijff-Korbayová, E. Oleari, A. Bagherzadhalimi, F. Sacchitelli, B. Kiefer, S. Racioppa, C. Pozzi, and A. Sanna, "Young users' perception of a social robot displaying familiarity and eliciting disclosure," In *International Conference on Social Robotics*, Paris, France, pp. 380–389, 2015.

[17] G. E. Birnbaum, M. Mizrahi, G. Hoffman, H. T. Reis, E. J. Finkel, and O. Sass, "What robots can teach us about intimacy: The reassuring effects of robot responsiveness to human disclosure," *Computers in Human Behavior*, vol. 63, pp. 416–423, 2016.

[18] N. Martelaro, V. C. Nneji, W. Ju, and P. Hinds, "Tell me more: Designing hri to encourage more trust, disclosure, and companionship," In *The 11th ACM/IEEE International Conference on Human Robot Interaction*, Christchurch, New Zealand, pp. 181–188, 2016.

[19] P. C. Cozby, "Self-disclosure: A literature review," *Psychological Bulletin*, vol. 79, no. 2, p. 73, 1973.

[20] J. L. Gibbs, N. B. Ellison, and R. D. Heino, "Self-presentation in online personals: The role of anticipated future interaction, self-disclosure, and perceived success in Internet dating," *Communication Research*, vol. 33, no. 2, pp. 152–177, 2006.

[21] S. M. Jourard, and J. E. Rubin, "Self-disclosure and touching: A study of two modes of interpersonal encounter and their inter-relation," *Journal of Humanistic Psychology*, vol. 8, no. 1, pp. 39–48, 1968.

[22] J. Weizenbaum, "ELIZA-a computer program for the study of natural language communication between man and machine," *Communications of the ACM*, vol. 9, no. 1, pp. 36–45, 1966.

[23] Y. Moon, "Intimate exchanges: Using computers to elicit self-disclosure from consumers," *Journal of Consumer Research*, vol. 26, no. 4, pp. 323–339, 2000.

[24] K. Suzuki, M. Yokoyama, Y. Kionshita, T. Mochizuki, T. Yamada, S. Sakurai, T. Narumi, T. Tanikawa, and M. Hirose, "Gender-impression modification enhances the effect of mediated social touch between persons of the same gender," *Augmented Human Research*, vol. 1, no. 1, pp. 1–11, 2016.

[25] M. Shiomi, K. Nakagawa, K. Shinozawa, R. Matsumura, H. Ishiguro, and N. Hagita, "Does a robot's touch encourage human effort?," *International Journal of Social Robotics*, vol. 9, pp. 5–15, 2016.

[26] X. Zheng, M. Shiomi, T. Minato, and H. Ishiguro, "What kinds of robot's touch will match expressed emotions?," *IEEE Robotics and Automation Letters*, pp. 127–134, 2019.

[27] X. Zheng, M. Shiomi, T. Minato, and H. Ishiguro, "How can robot make people feel intimacy through touch?," *Journal of Robotics and Mechatronics*, vol. 32, no. 1, pp. (to appeear), 2019.

[28] X. Zheng, M. Shiomi, T. Minato, and H. Ishiguro, "Modeling the timing and duration of grip behavior to express emotions for a social robot," *IEEE Robotics and Automation Letters*, vol. 6, no. 1, pp. 159–166, 2020.

[29] M. Shiomi, X. Zheng, T. Minato, and H. Ishiguro, "Implementation and evaluation of a grip behavior model to express emotions for an android robot," *Frontiers in Robotics and AI*, vol. 8, 2021.

[30] M. Shiomi, A. Nakata, M. Kanbara, and N. Hagita, "Robot reciprocation of hugs increases both interacting times and self-disclosures," *International Journal of Social Robotics*, pp. 1–9, 2020.

[31] M. Shiomi, A. Nakata, M. Kanbara, and N. Hagita, "A robot that encourages self-disclosure by hug," In *Social Robotics: 9th International Conference, ICSR 2017, Tsukuba, Japan, November 22–24, 2017, Proceedings*, A. Kheddar, E. Yoshida, S. S. Ge et al., eds., pp. 324–333, Cham: Springer International Publishing, 2017.

[32] N. Dahlbäck, A. Jönsson, and L. Ahrenberg, "Wizard of Oz studies: Why and how," In *Proceedings of the 1st International Conference on Intelligent User Interfaces*, Orlando, Florida, USA, pp. 193–200, 1993.

[33] H. Sumioka, A. Nakae, R. Kanai, and H. Ishiguro, "Huggable communication medium decreases cortisol levels," *Scientific Reports*, vol. 3, p. 3034, 2013.

[34] R. Yu, E. Hui, J. Lee, D. Poon, A. Ng, K. Sit, K. Ip, F. Yeung, M. Wong, and T. Shibata, "Use of a therapeutic, socially assistive pet robot (PARO) in improving mood and stimulating social interaction and communication for people with dementia: Study protocol for a randomized controlled trial," *JMIR Research Protocols*, vol. 4, no. 2, p. e4189, 2015.

[35] K. Nakagawa, M. Shiomi, K. Shinozawa, R. Matsumura, H. Ishiguro, and N. Hagita, "Effect of robot's whispering behavior on people's motivation," *International Journal of Social Robotics*, vol. 5, no. 1, pp. 5–16, 2012.

[36] H. Fukuda, M. Shiomi, K. Nakagawa, and K. Ueda, "'Midas touch' in human-robot interaction: Evidence from event-related potentials during the ultimatum game," In *2012 7th ACM/IEEE International Conference on Human-Robot Interaction (HRI)*, pp. 131–132, 2012.

[37] M. Shiomi, A. Nakata, M. Kanbara, and N. Hagita, "A hug from a robot encourages prosocial behavior," In *2017 26th IEEE International Symposium on Robot and Human Interactive Communication (RO-MAN)*, Lisbon, Portugal, pp. 418–423, 2017.

[38] M. Lombard, J. Snyder-Duch, and C. C. Bracken, "Content analysis in mass communication: Assessment and reporting of intercoder reliability," *Human Communication Research*, vol. 28, no. 4, pp. 587–604, 2002.

[39] J. Cohen, "A coefficient of agreement for nominal scales," *Educational and Psychological Measurement*, vol. 20, no. 1, pp. 37–46, 1960.

[40] G. Huisman, "Social touch technology: A survey of haptic technology for social touch," *IEEE Transactions on Haptics*, vol. 10, no. 3, pp. 391–408, 2017.

[41] G. M. Lucas, J. Gratch, A. King, and L.-P. Morency, "It's only a computer: Virtual humans increase willingness to disclose," *Computers in Human Behavior*, vol. 37, pp. 94–100, 2014.

CHAPTER 15

Audio-Visual Stimuli Improve Both Robot's Hug Impressions and Stress-Buffering Effects

Masahiro Shiomi

Advanced Telecommunications Research Institute International, Kyoto, Japan

Norihiro Hagita

Osaka University of Arts, Osaka, Japan

15.1 INTRODUCTION

The human–robot interaction field has witnessed rising interest in tactile engagement, prompting researchers to focus on the study of touch in social robots. Fueled by advancements in technology, robots are increasingly engaged in physical interactions with humans, such as handshakes and hugs, in various environments, including elementary schools [1,2], museums [3,4], and shopping centers [5,6]. Amicable interactions and beneficial outcomes have been demonstrated in human–human touch interaction, highlighting the importance of touch in fostering positive relationships [7–12]. Following these phenomena, robotics researchers have investigated human–robot tactile interactions in diverse contexts, including stress

DOI: 10.1201/9781003384274-20

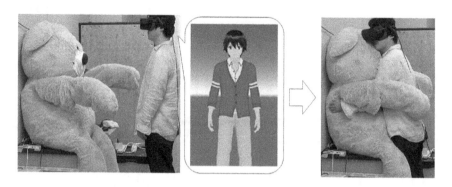

FIGURE 15.1   MetaHug system.

buffering [7], motivation improvement [8], mental health support [9], and the promotion of prosocial behavior [10].

Literature from the field of human sciences indicates that gender plays a role in shaping the perception of touch between individuals [11,12]. However, previous studies have primarily centered on robots with machine-like or pet-like appearances, overlooking the potential impact of a robot's perceived gender on touch interactions. Thus, it remains unclear how a robot's perceived gender might influence the impressions of its hug. Investigating the role of gender in robot touch interactions is fraught with challenges such as hardware configuration limitations and financial constraints.

To circumvent these obstacles, we utilized a huggable robot in conjunction with a virtual reality (VR) application, which facilitated the simple adjustment of its appearance and voice through audio-visual stimuli. Employing the MetaHug system [13] (Figure 15.1), our research, which explored how altering audio-visual stimuli in a VR application changes a robot's perceived gender, enables us to investigate how such changes impact human participants' perceptions of a robot's hug.

In addition to these inquiries, our study delves into the potential stress-buffering effects associated with a robot's hug. Prior research in the field of human sciences has identified stress reduction that results from actual tactile interaction (including hugs) with actual intimate [14] and imagined tactile interactions [15]. Although stress-buffering effects have been observed in human–robot tactile interactions [7], the influence of perceived gender has not been explored. Understanding the impact of a robot's perceived gender on stress buffering might inform the design of future touch interactions between humans and robots and shape the development of mental support robotic systems. Thus, we investigate how the perceived gender of a robot changes the stress-buffering effects of its hugs. Note that this chapter is modified based on our previous work [16], edited to be comprehensive and fit with the context of this book.

15.2 SYSTEM

The MetaHug system includes a motion control component that oversees a VR application (i.e., virtual agent movements and vocalizations) and a robot system (i.e., robot movements).

15.2.1 VR Application

We employed Unity and Oculus Rift to manage the audio-visual stimuli for participants through a VR application. We tracked their head positions using two Oculus sensors, which monitored the Oculus Rift's position. We also prepared virtual agents with masculine and feminine appearances with voice synthesis functions. We standardized their heights to eliminate any size-related biases. We implemented autonomous eye contact functions with users based on head position data, and a function to control their lip movements that adjusts for synchronization with speech content. We also designed hugging animations for both agents.

15.2.2 Robot

We used Moffuly [10], a robot capable of hugging people. It has a single degree of freedom (DOF) for each elbow and adequate arm length for hugging interactions. A touch sensor (ShokacCube, developed by Touchence) was integrated into the ends of its arms for identifying contact with individuals to ensure their safety; if a certain pressure level is detected, the hugging motion stops, and Moffuly slightly opens its arm.

15.2.3 Motion Controller

The motion control component utilizes sensor data from the VR application and the robot system (i.e., head position and pressure information) to coordinate the movements, the vocalizations of the virtual agents, and the robot's motions for the synchronized hugs. The motion controller initiates hugging movements based on the user's head position. During an interaction, the robot gently pats the user's back. A one-minute hugging duration was designed so that both the virtual agents and the robot simultaneously ensure that their hugging movements conclude. Although prior human science studies demonstrated that intimate hugs provide numerous positive effects, they did not specify an ideal hug length; therefore, we developed a relatively long hugging behavior (i.e., one minute) to convey a sense of intimacy.

15.3 EXPERIMENT

15.3.1 Hypothesis and Predictions

As discussed in Section 15.2, the perception of gender plays a significant role in touch interactions between individuals [11]. For instance, previous research argued that touch from a person of the opposite gender is generally well-received, although touch from a person of the same gender can produce mixed reactions, including both acceptance and rejection [11, 12]. Thus, if the MetaHug system effectively manipulates the perceived gender of virtual agents, users are likely to have a more favorable response to hugging experiences when interacting with an agent of the gender of romantic interest. With this in mind, we propose the following prediction regarding hugging impressions:

Prediction 1: Participants will experience increased comfort and a higher willingness for another hug when interacting with an agent of the gender of romantic interest.

As mentioned in Section 15.2, previous research demonstrated that hugging interactions offer stress-buffering effects, regardless of whether the interaction partner is a person actual intimate [14,17], an imagined intimate [15], or a robot [7].

However, these studies failed to investigate the potential impact of perceived gender, which we believe probably influences stress-buffering effects by hugging interactions, similar to other touch interactions [11,12]. Based on this reasoning, we formed a prediction about stress buffering.

Prediction 2: Hugging interactions with an agent of the gender of romantic interest will lead to greater reduction in participants' stress levels.

15.3.2 Participants

Eighteen Japanese individuals participated in our experiment, consisting of nine women and nine men who self-identified their genders. Their average age was 36.25, with a standard deviation of 8.74.

15.3.3 Conditions

Our study employed a mixed factorial design, combining within- and between-participant designs. We counterbalanced the order of the two within-participant conditions: same-gender and opposite-gender. The participant's gender served as the between-participant factor.

Same-gender condition: Participants engaged with an agent of the same gender.

Opposite-gender condition: Participants engaged with an agent of the opposite gender.

15.3.4 Procedures

Initially, we provided a brief overview of the experiment's objectives and procedures. After obtaining signed consent, we explained the system and demonstrated how to use it for hugging while wearing an HMD. In both conditions, the agents verbally requested a hug, and then they hugged the participants to facilitate the experience and expressed gratitude for their involvement in the experiment. Following the hugging interaction, the participants completed a subtraction task as a stress-inducing activity, a method commonly employed in human sciences literature to reliably induce stress [18,19]. We prepared three subtraction tasks (2,091 to 0 in 17-step sequences, 2,337 to 0 in 19-step sequences, or 3,567 to 0 in 29-step sequences) and counterbalanced their order. Each task lasted for five minutes. Participants were instructed to calculate as quickly as possible and warned that if they made errors, they would need to start over.

15.3.5 Measurements

To measure the participants' feelings about the hugging interaction, we designed a two-item questionnaire: (1) willingness to receive another hug and (2) comfort level during the hug experience, rated on a 1-to-7 point scale, with 7 representing the most favorable response.

To measure the participants' stress levels throughout the task, we used a technique from a previous study [15] in which participants self-reported their perceived stress while performing the tasks.

This technique was chosen because previous findings suggested that it provided a more accurate measure of perceived stress during the tasks, as opposed to ratings taken after task completion. Self-rated stress levels also showed a temporal relationship between physiological and psychological indicators of stress [20]. Participants reported their stress ratings (on a 0-to-10 scale, where 0 is no stress and 10 is extremely stressed) at 30-second intervals (prompted by a tone) during each five-minute task. We collected ten stress ratings for each condition, resulting in a Cronbach's α value of 0.97.

15.4 RESULTS

15.4.1 Verification of Prediction 1

Figure 15.2 shows mean and *SE* for comfort impressions. We performed a two-way mixed ANOVA with two factors: participant-gender (between-participant factor) and agent-gender (within-participant factor). The results showed significant differences in the agent-gender factor (F (1, 14)=12.962, p=0.002, η^2=0.448) and in the interaction effect (F (1, 14)=12.962, p=0.002, η^2=0.448). They did not show any significant differences in the participant-gender factor (F (1, 14)=0.899, p=0.357, η^2=0.053). We conducted a multiple comparison with the Bonferroni method, which showed a significant difference for the same-gender agent: women>men (p=0.043). For the opposite-gender agent, we found no significant difference between the women and men (p=0.594). The results also showed a significant difference for men: opposite-gender agents>same-gender agents (p<0.001). For women, we found no significant difference between opposite- and same-gender agents (p=1.00).

Figure 15.3 shows the mean and *SE* for willingness for another hug. We conducted a two-way mixed ANOVA with two factors: participant-gender (between-participant) and agent-gender (within-participant factor). The results showed significant differences in the agent-gender factor (F (1, 14)=5.982, p=0.026, η^2=0.272) and in the interaction effect

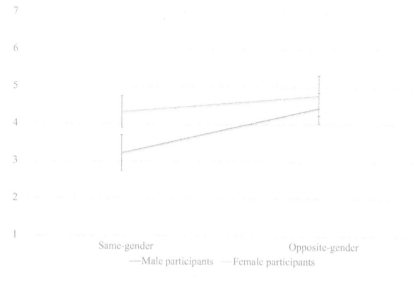

FIGURE 15.2 Questionnaire results about comfortableness of hug interaction.

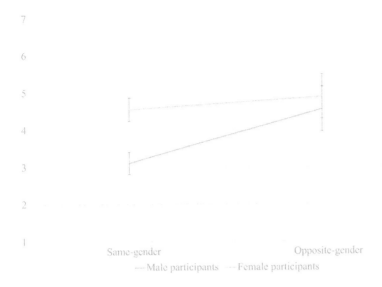

FIGURE 15.3 Questionnaire results about willingness for another hug.

$(F$ (1, 14)$=7.965$, $p=0.012$, $\eta^2=0.332$), but not in the participant-gender factor $(F$ (1, 14)$=0.429$, $p=0.522$, $\eta^2=0.026$). We conducted a multiple comparison with the Bonferroni method, which did not show any significant difference for either the opposite-gender agent $(p=0.466)$ or the same-gender agent $(p=0.111)$. On the other hand, the results showed a significant difference for men: opposite-gender agents > same-gender agents $(p=0.002)$. But for women, we found no significant difference $(p=0.794)$.

The finding offered partial support for prediction 1. A notable increase in positive impressions concerning comfort and desire for an additional hug was observed solely among men when interacting with an opposite-gender agent as opposed to a same-gender agent.

15.4.2 Verification of Prediction 2

Figure 15.4 shows the mean and *SE* for perceived stress. We performed a two-way mixed ANOVA with two factors: participant-gender (between-participant factor) and agent-gender (within-participant factor). The results showed significant differences in the agent-gender factor $(F$ (1, 16)$=4.768$, $p=0.044$, $\eta^2=0.230$). We did not find any significant differences in the gender factor $(F$ (1, 16)$=0.459$, $p=0.508$, $\eta^2=0.028$) or in the interaction effect $(F$ (1, 16)$=2.700$, $p=0.120$, $\eta^2=0.120$). These findings suggest that for women and men, engaging in a hug interaction with

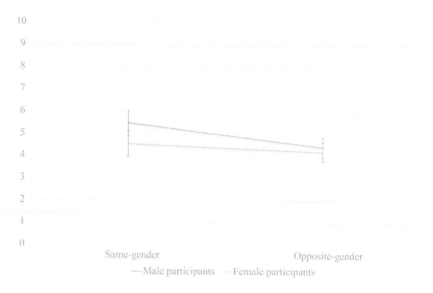

FIGURE 15.4 Questionnaire results about perceived stress during tasks (0 means no stress).

an opposite-gender agent was more effective in a stress-buffering context than with a same-gender agent. Consequently, prediction 2 was supported.

15.4.3 Additional Analysis of Task Performances

We investigated the stress-buffering effects and impressions of hug interactions by manipulating the perceived gender of the interacting agents. To deepen our understanding of the impact of hugs, we explored how perceived gender during hug interactions influenced the participants' task performance.

We measured their performance scores, as represented by the number of accurately completed serial subtractions (Figure 15.5). A two-way mixed ANOVA was conducted with two factors: participant-gender (a between-participant factor) and agent-gender (a within-participant factor). This approach allowed us to gain deeper insights into the relationship between perceived gender during hug interactions and the performance of the participants. The results did not show any significant differences for any of the factors: the agent-gender factor (F (1, 16)=0.01, p=0.980, η^2=0.001), the participant-gender factor (F (1, 16)=1.304, p=0.270, η^2=0.075), or the interaction effect (F (1, 16)=0.337, p=0.569, η^2=0.021). Thus, the perceived gender did not significantly affect the task performances.

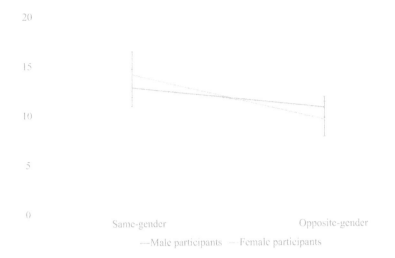

FIGURE 15.5   Task performance.

15.5 DISCUSSION

15.5.1 Implications

Our findings indicate that both women and men experience stress-buffering effects during hug interactions with opposite-gender agents. Interestingly, the agent-gender factor significantly influenced hug impressions only for men. Although some of our hypotheses were not supported, the results suggest that hug interactions with an opposite-gender agent generally provide greater benefits than those with a same-gender agent.

Another key insight derived from our study is the similarity between human–human and human–robot touch interactions in terms of observed trends. Previous research has demonstrated that women tend to respond more positively to touch interactions than men [11], and touches between men can have negative effects [12]. The promising outcomes of our study, which involved hug interactions with a robot and a VR application to manipulate the perceived gender of the interacting agents, suggest that this approach holds significant potential. By allowing for simple modifications to the agents' appearances and voices, this method offers advantages over traditional robot designs.

15.5.2 Limitations

Our study has several limitations. First, we used a one-minute hug, which we chose to represent an intimate hug. We should explore the effects of different hug lengths or intra-hug gestures [21–24] to determine the minimum time needed for a hug to reduce stress. In particular, past studies reported what kinds of touch characteristics are important to convey emotions and intimate feelings [8,25–31]; such knowledge will improve our stress-buffering hug interaction design. We also conducted our experiments in a lab setting. Our system must be tested in real-life situations. To do this, we need to create a natural setting for hug interactions and develop a suitable conversation design, which our study did not address.

Second, our participant pool lacks diversity in terms of age, cultural background, or prior experience with robots, limiting the generalizability of our findings. Future research should include a more diverse group of participants to understand better the effects of perceived gender and hug interactions across different populations.

Third, the virtual agents used in the study might not represent the full spectrum of possible agent designs, including variations in appearance, voice, and behavior. Exploring a broader range of virtual agent characteristics in future studies will help determine how different aspects of agent design influence the outcomes of hug interactions. Related to this topic, using more realistic and human-like robots [32,33] without virtual agents will provide rich knowledge about hug interaction effects.

15.6 CONCLUSION

We examined the influence of virtual agents' perceived gender on the impressions and stress-buffering effects of hug interactions. We conducted experiments with participants using the MetaHug system, which combines a huggable robot and a VR application to facilitate physical hug interactions between people and virtual agents. They engaged in two hug interactions with distinct virtual agents, completed stressful tasks, and subsequently reported their perceived stress levels and impressions of the hugs. Our findings revealed that, regardless of the participants' gender, hug interactions with opposite-gender agents yielded stress-buffering effects. However, only men experienced an improvement in their hug impressions. This study sheds light on the significance of perceived gender in hug interactions with virtual agents and its potential implications for future applications.

ACKNOWLEDGMENT

This work was supported by JST CREST Grant Number JPMJCR18A1, Japan.

REFERENCES

[1] T. Kanda, R. Sato, N. Saiwaki, and H. Ishiguro, "A two-month field trial in an elementary school for long-term human-robot interaction," *IEEE Transactions on Robotics*, vol. 23, no. 5, pp. 962–971, 2007.

[2] M. Shiomi, T. Kanda, I. Howley, K. Hayashi, and N. Hagita, "Can a social robot stimulate science curiosity in classrooms?," *International Journal of Social Robotics*, vol. 7, no. 5, pp. 641–652, 2015.

[3] I. R. Nourbakhsh, C. Kunz, and T. Willeke, "The mobot museum robot installations: A five year experiment," In *Proceedings of the 2003 IEEE/ RSJ International Conference on Intelligent Robots and Systems*, Las Vegas, Nevada, United States, pp. 3636–3641, 2003.

[4] M. Shiomi, T. Kanda, H. Ishiguro, and N. Hagita, "Interactive humanoid robots for a science museum," *IEEE Intelligent Systems*, vol. 22, no. 2, pp. 25–32, 2007.

[5] H.-M. Gross, H.-J. Böhme, C. Schröter, S. Mueller, A. König, C. Martin, M. Merten, and A. Bley, "Shopbot: Progress in developing an interactive mobile shopping assistant for everyday use," In *SMC 2008 IEEE International Conference on Systems, Man and Cybernetics*, Singapore, Singapore, pp. 3471–3478, 2008.

[6] S. Satake, K. Hayashi, K. Nakatani, and T. Kanda, "Field trial of an information-providing robot in a shopping mall," In *2015 IEEE/RSJ International Conference on Intelligent Robots and Systems* (IROS), Hamburg, Germany, pp. 1832–1839, 2015.

[7] H. Sumioka, A. Nakae, R. Kanai, and H. Ishiguro, "Huggable communication medium decreases cortisol levels," *Scientific Reports*, vol. 3, p. 3034, 2013.

[8] M. Shiomi, K. Nakagawa, K. Shinozawa, R. Matsumura, H. Ishiguro, and N. Hagita, "Does a robot's touch encourage human effort?," *International Journal of Social Robotics*, vol. 9, pp. 5–15, 2016.

[9] R. Yu, E. Hui, J. Lee, D. Poon, A. Ng, K. Sit, K. Ip, F. Yeung, M. Wong, and T. Shibata, "Use of a therapeutic, socially assistive pet robot (PARO) in improving mood and stimulating social interaction and communication for people with dementia: Study protocol for a randomized controlled trial," *JMIR Research Protocols*, vol. 4, no. 2, p. e4189, 2015.

[10] M. Shiomi, A. Nakata, M. Kanbara, and N. Hagita, "A hug from a robot encourages prosocial behavior," In *2017 26th IEEE International Symposium on Robot and Human Interactive Communication (RO-MAN)*, Lisbon, Portugal, pp. 418–423, 2017.

[11] D. S. Stier, and J. A. Hall, "Gender differences in touch: An empirical and theoretical review," *Journal of Personality and Social Psychology*, vol. 47, no. 2, p. 440, 1984.

[12] J. A. Evans, "Cautious caregivers: Gender stereotypes and the sexualization of men nurses' touch," *Journal of Advanced Nursing*, vol. 40, no. 4, pp. 441–448, 2002.

[13] M. Shiomi, and N. Hagita, "Do audio-visual stimuli change hug impressions?," In *Social Robotics: 9th International Conference, ICSR 2017, Tsukuba, Japan, November 22–24, 2017, Proceedings*, A. Kheddar, E. Yoshida, S. S. Ge et al., eds., pp. 345–354, Cham: Springer International Publishing, 2017.

[14] S. Cohen, D. Janicki-Deverts, R. B. Turner, and W. J. Doyle, "Does hugging provide stress-buffering social support? A study of susceptibility to upper respiratory infection and illness," *Psychological Science*, vol. 26, no. 2, pp. 135–147, 2015.

[15] B. K. Jakubiak, and B. C. Feeney, "Keep in touch: The effects of imagined touch support on stress and exploration," *Journal of Experimental Social Psychology*, vol. 65, pp. 59–67, 2016.

[16] M. Shiomi, and N. Hagita, "Audio-visual stimuli change not only robot's hug impressions but also its stress-buffering effects," *International Journal of Social Robotics*, pp. 1–8, 2019.

[17] K. M. Grewen, B. J. Anderson, S. S. Girdler, and K. C. Light, "Warm partner contact is related to lower cardiovascular reactivity," *Behavioral Medicine*, vol. 29, no. 3, pp. 123–130, 2003.

[18] M. A. Birkett, "The trier social stress test protocol for inducing psychological stress," *Journal of Visualized Experiments*, no. 56, 2011.

[19] J. D. Creswell, W. T. Welch, S. E. Taylor, D. K. Sherman, T. L. Gruenewald, and T. Mann, "Affirmation of personal values buffers neuroendocrine and psychological stress responses," *Psychological Science*, vol. 16, no. 11, pp. 846–851, 2005.

[20] J. Hellhammer, and M. Schubert, "The physiological response to trier social stress test relates to subjective measures of stress during but not before or after the test," *Psychoneuroendocrinology*, vol. 37, no. 1, pp. 119–124, 2012.

[21] A. E. Block, S. Christen, R. Gassert, O. Hilliges, and K. J. Kuchenbecker, "The six hug commandments: Design and evaluation of a human-sized hugging robot with visual and haptic perception," In *Proceedings of the 2021 ACM/IEEE International Conference on Human-Robot Interaction*, pp. 380–388, 2021.

[22] A. E. Block, H. Seifi, O. Hilliges, R. Gassert, and K. J. Kuchenbecker, "In the arms of a robot: Designing autonomous hugging robots with intra-hug gestures," *ACM Transactions on Human-Robot Interaction*, vol. 12, no. 2, pp. 1–49, 2022.

[23] Y. Onishi, H. Sumioka, and M. Shiomi, "Increasing torso contact: Comparing human-human relationships and situations," In *International Conference on Social Robotics*, Singapore, Singapore, pp. 616–625, 2021.

[24] Y. Onishi, H. Sumioka, and M. Shiomi, "Designing a robot which touches the user's head with intra-hug gestures," In *Companion of the 2023 ACM/IEEE International Conference on Human-Robot Interaction*, Stockholm, Sweden, pp. 314–317, 2023.

[25] B. Alenljung, R. Andreasson, R. Lowe, E. Billing, and J. Lindblom, "Conveying emotions by touch to the nao robot: A user experience perspective," *Multimodal Technologies and Interaction*, vol. 2, no. 4, p. 82, 2018.

[26] X. Zheng, M. Shiomi, T. Minato, and H. Ishiguro, "What kinds of robot's touch will match expressed emotions?," *IEEE Robotics and Automation Letters*, vol. 5, pp. 127–134, 2019.

[27] X. Zheng, M. Shiomi, T. Minato, and H. Ishiguro, "Modeling the timing and duration of grip behavior to express emotions for a social robot," *IEEE Robotics and Automation Letters*, vol. 6, no. 1, pp. 159–166, 2020.

[28] M. Shiomi, X. Zheng, T. Minato, and H. Ishiguro, "Implementation and evaluation of a grip behavior model to express emotions for an android robot," *Frontiers in Robotics and AI*, vol. 8, p. 755150, 2021.

[29] X. Zheng, M. Shiomi, T. Minato, and H. Ishiguro, "How can robot make people feel intimacy through touch?," *Journal of Robotics and Mechatronics*, vol. 32, no. 1, pp. 51–58, 2020.

[30] K. Suzuki, M. Yokoyama, Y. Kinoshita, T. Mochizuki, T. Yamada, and S. Sakurai, "Enhancing effect of mediated social touch between same gender by changing gender impression," In *Proceedings of the 7th Augmented Human International Conference 2016*, Geneva, Switzerland, pp. 1–8, 2016.

[31] K. Higashino, M. Kimoto, T. Iio, K. Shimohara, and M. Shiomi, "Tactile stimulus is essential to increase motivation for touch interaction in virtual environment," *Advanced Robotics*, vol. 35, no. 17, pp. 1043–1053, 2021.

[32] D. F. Glas, T. Minato, C. T. Ishi, T. Kawahara, and H. Ishiguro, "Erica: The erato intelligent conversational android," In *2016 25th IEEE International Symposium on Robot and Human Interactive Communication (RO-MAN)*, pp. 22–29, 2016.

[33] M. Shiomi, H. Sumioka, K. Sakai, T. Funayama, and T. Minato, "SŌTO: An android platform with a masculine appearance for social touch interaction," In *Companion of the 2020 ACM/IEEE International Conference on Human-Robot Interaction*, Cambridge, United Kingdom, pp. 447–449, 2020.

CHAPTER 16

Praise with Tactile Stimulus Increases Motivation

Higashino Kana, Mitsuhiko Kimoto,
Takamasa Iio, Katsunori Shimohara,
and Masahiro Shiomi

*Advanced Telecommunications Research
Institute International, Kyoto, Japan*

Doshisha University, Kyoto, Japan

16.1 INTRODUCTION

The impact of physical touch on individuals' behaviors and perceptions is a significant aspect of human–robot interaction [1,2]. Previous research has identified the following positive aspects or results when a robot engages in touch: a fostering of motivation [3], increased persuasive power [4], and encouraged pro-social behaviors [5]; a sharing of personal information by self-disclosures [6]; increased pain- or stress-buffering effects [7,8]; and conveying various emotions [9–12]. By utilizing tactile interaction, robots have the ability to leave positive impressions with interactive people [13] and express a variety of emotions through adjustments in touch properties [9,14]. These studies provide rich knowledge about the positive effects of human–robot touch interaction.

However, the necessity of actual touching remains an open question, as recent innovations in virtual reality (VR) applications have introduced

DOI: 10.1201/9781003384274-21

interactions with others using pseudo-haptic stimuli, such as visually touching behavior. A recent study concerning VR applications demonstrated the effectiveness of such virtual-touch interaction within VR environments, even though it relied solely on visual stimuli [15]. In contrast, research on body-transfer illusion effects, like the rubber hand illusion, has probed the influence of visual and tactile stimuli on individuals' perceptions [16–21]. Yet it has paid less attention to the positive outcomes of touch behaviors in interactions between robots and people-based interactions with others.

Based on these considerations, we focused on pseudo-touch interactions, particularly those limited to visual-only touch, which might produce comparable modifications in individual behaviors and perceptions. To investigate such effects, we experimentally compared the motivational enhancement effects between visual-only touch and a combination of visual-tactile-touch interactions by integrating a VR application with a physical robot (Figure 16.1). Note that this chapter is modified based on our previous work [22], edited to be comprehensive and fit with the context of this book.

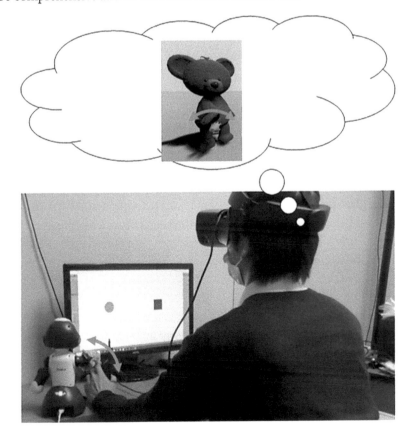

FIGURE 16.1   Touch interaction with a virtual agent.

16.2 SYSTEM

Our system consists of four distinct hardware components: a computer, a head-mounted display, a touch controller, and a robot. The computer executes a VR environment application to manage both the agent and the robot based on an experimental scenario. We utilized an Oculus Rift S head-mounted display for the experiment, and an Oculus Touch controller tracked the positions of the left hands of the users.

Our system contains eight distinct software components. First, the scenario manager loads a pre-designed experimental procedure containing both verbal and non-verbal actions for the agent and conveys the particulars of the agent's actions to the behavior interpreter, which subsequently translates them for the agent controller. If a touch-related command is present, it is also forwarded to the robot controller to manage the robot's arm movement. The agent controller collaborates with the VR world manager to modify the VR landscape's state. Next, the visual/audio renderer displays the updated state. The modules that sense the head and hand positions gather the users' head and hand positions for the behavior interpreter, facilitating the agent's behavior modification. This position information allows the agent to perform eye contact behaviors and refine its touch motion as necessary.

16.2.1 VR Application

Our system, which uses a 3D virtual agent that resembles a bear-like character, carries out three distinct movement categories: idling, scenario-based, and touch-based behaviors. Idle actions and eye contact fall under idling behaviors in which the agent gently sways its head to engage users in the VR space. The scenario-based behaviors feature clapping and waving. The touch-based behaviors, also specified in the scenario, are adapted to connect with a user's left hand by sensing modules. We created pre-recorded speech for the agent with speech synthesis software.

16.2.2 Robot

The experiment employed Sota, a tabletop-sized robot that possesses eight degrees of freedom (DOFs) distributed among its head (three), arms (two each), and lower body (one). Sota stands 28 cm tall. Although a humanoid robot is not essential for tactile stimulus delivery, we chose Sota due to its simple operation. Since the robot's sole purpose in this experiment is making contact with a user's left hand, we covered its right hand in plush fabric. The robot's touch movement was synchronized with the corresponding action performed by the virtual agent

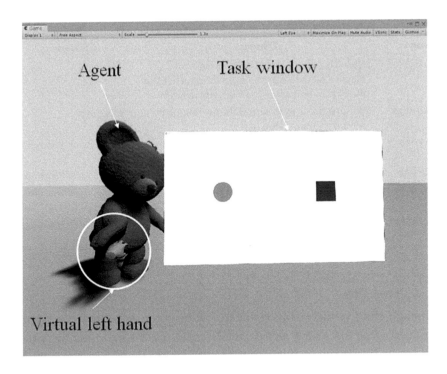

FIGURE 16.2   Image seen in practice sessions.

16.3 SCENARIO

16.3.1 Practice

At the experiment's beginning, the experimenter explained this study's procedure to the participants who earlier provided written informed consent. Next, they used the VR headset (viewpoint is shown in Figure 16.2) and listened to detailed procedures from the virtual agent. The participants could see the task window, the virtual agent, and their own virtual left hands. They engaged in several practice sessions of the task. After completing them, the agent explained the next phase, the fixed-time session, whose details are described in the next subsection. Table 16.1 showed utterance contents of the agent in the experiment.

16.3.2 Fixed Time

The fixed-time session's duration lasted five minutes. First, the agent asked the participants to continue with the task throughout the session while informing them of its length. Every 30 seconds, the agent praised the participants, incorporating three different motion types. After completing the five-minute session, the agent again explained the next phase, the free-time session, whose details are described in the next subsection.

TABLE 16.1 Scenario Contents

Session	Utterance	Motion
Practice	Hello. My name is Teddy. Nice to meet you. Please read the instructions on the screen. Click the start button when you are ready to begin.	Waving
Fixed-time	Your practice is over. Next is the five-minute, second session. Good luck.	Clapping
After 30 s	You're off to a good start. You're getting faster than in your practice.	Clapping or Touching
After 60 s	One minute has passed. Keep working.	Clapping or Touching
After 90 s	You're working on your tasks at a good pace. Keep it up.	Clapping or Touching
After 120 s	You seem to be getting used to it. You did more tasks than most other participants.	Clapping or Touching

16.3.3 Free Time

Although the free-time sessions lasted 15 minutes, note that the agent refrained from disclosing any particular maximum time, thus the participants were not aware of any time limitations. At the session's start, the agent once again requested them to continue to perform the task without explicitly mentioning a time frame. The agent did clarify that they could terminate the task at any time. During this session, the agent maintained a distance and neither offered praise nor initiated touch, different from the fixed-time session. The free-time session finished when the participants pressed the ESC key or reached the maximum time limit.

16.3.4 Task

We utilized a repetitive and monotonous drag-and-drop task that previously examined touch effects in human–robot experiments [3]. A circle and a square are displayed on the headset's screen, and users drag the circle into the square. After successful completion, the circle vanishes, and a new one emerges in its original position. Users repeatedly performed this operation, and the mouse cursor's speed was deliberately set low.

16.4 EXPERIMENT

16.4.1 Hypotheses and Predictions

Prior research has underscored the role of touch interactions with robots for boosting motivation [3]. Unfortunately, the effects of visual and tactile stimuli in touch interaction remain inadequately investigated since such an approach is deemed beyond the scope of research. Whether tactile stimuli actually enhance motivational effects remains ambiguous. This research gap prompts a simple yet essential question: How vital is the presence of tactile stimuli for improving motivation by touch interactions? Although earlier studies that investigated the effects of visual-only touch did not focus on motivation enhancement, they did reveal changes in people's perceptions [15,23–25]. Another study discovered that imagined touch, for instance, without tactile stimuli, had favorable effects on stress buffering [26].

To tackle this question, we designed an experiment that compared the outcomes of visual-only and visual-tactile touch in a virtual setting. We hypothesized that the latter will yield greater benefits than the former since previous research indicated that the activation of C-tactile fibers at a suitable speed (5–10 cm/s) evokes pleasant feelings [27].

These sensations provide positive impressions and modify people's behaviors [3]. Based on these hypotheses, we made the following predictions:

Prediction 1: Visual-tactile touches will encourage the participants to do more tasks than the visual-only touches.

Prediction 2: Visual-tactile touches will create more positive feelings in the participants toward the robot than the visual-only touches.

16.4.2 Conditions

We employed a between-participants design to analyze and compare the effects of two different touch conditions: visual-only and visual-tactile. The following are the details of these conditions:

- **Visual-only-touch condition:** The virtual agent interacts with the participants by touching their virtual left hands without any tactile feedback.

- **Visual-tactile-touch condition:** The virtual agent interacts with the participants by touching their virtual left hands; the robot also physically touches their actual left hands, providing tactile sensations.

We decided not to incorporate a physical touch-only condition in which the robot physically touches the participants' hands without any virtual interaction from the agent due to potential discomfort and an unnatural disconnection between the visual and tactile interactions.

In this study, we did not investigate the gender factor because a previous analysis found [28] no significant gender differences. It also demonstrated that both the no-touch and visual-only-touch conditions yielded similar patterns when compared with the visual-tactile-touch condition. Unlike the influence of gender, excluding both conditions hinders the possibility of examining the effects of visual-tactile-touch. Since the main focus of our study is investigating the effects of tactile stimuli in a praise context, we did not include a no-touch condition and retained the visual-only-touch condition as an alternative.

16.4.3 Participants

Our 48 participants were comprised of an equal number of women and men whose ages ranged from 21 to 54, with an average of 35.7 ($SD = 11.0$). For each condition, we allocated 12 women and 12 men, i.e., 24 participants. Unfortunately, we excluded four participants due to misinterpretations of instructions and disruptions in each condition. We obtained valid results from 11 women and nine men in the visual-only-touch condition and from nine women and 11 men in the visual-tactile condition.

16.4.4 Procedure

As described in Section 16.3.1, at the beginning of the experiment, participants were provided with an overview of its objectives and procedures. This research was approved by our institution's ethics committee for studies involving human subjects, and written informed consent was obtained from every participant. The experimenter asked the participants to put on their headsets. Next, the experimenter positioned the robot in the visual-tactile condition and left the room. Following the system setup, the system showed instructions on the virtual environment screen. During the practice session, they performed the task several times, repeated it for five minutes in the fixed-time session, and resumed it during the free-time session until they chose to stop or the 15-minute time limit was reached.

Afterward, the experimenter re-entered the room, concealed the robot from the participants, asked them to remove their headsets, and handed out questionnaires. At the experiment's conclusion, the experimenter informed the participants in the visual-tactile-touch condition how the tactile stimuli were delivered.

16.4.5 Measurements

We measured the task motivation by following a subjective item: the duration of the free-time sessions. We used an objective item to evaluate the participants' impressions of the agent with a likeability scale [29].

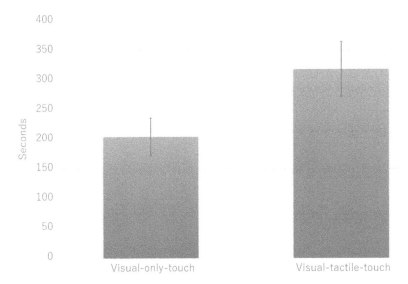

FIGURE 16.3   Working time in free-time sessions.

16.5 RESULTS

16.5.1 Verification of Prediction 1

Figure 16.3 shows the duration of the free-time session. We conducted a t-test whose results showed significant differences between the conditions ($t(38)=2.027$, $p=0.050$). Thus, the participants in the *visual-tactile-touch* condition did significantly more tasks than the participants in the *visual-only-touch* condition; prediction 1 was supported.

16.5.2 Verification of Prediction 2

Figure 16.4 shows the perceived likeability of the participants in each group. We conducted a t-test whose results did not show significant differences between conditions ($t(38)=0.555$, $p=0.582$). Thus, the visual-tactile touch did not increase the participants' positive impressions compared to the visual-only-touch; prediction 2 was not supported.

16.6 DISCUSSION

The experiment results showed several implications. First, our results highlight the significance of tactile stimuli for enhancing motivation through touch interaction. They demonstrate that the participants in the visual-tactile-touch condition engaged in tasks for a longer duration compared to those in the visual-only-touch condition. These findings suggest a

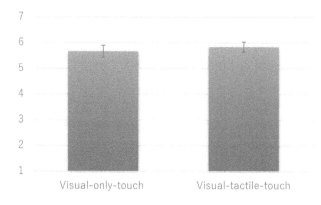

FIGURE 16.4   Perceived likeability.

promising possibility for employing robotic devices to deliver tactile stimuli in VR applications, such as education and rehabilitation.

One benefit of integrating these applications with physical devices is the ability to bypass their appearance effects. Appearances influence touch effects. For example, a past study developed an application that integrates a virtual agent and a physical robot to investigate visual-audio tactile stimuli effects in the context of stress-buffering effects [8]. Our results also provide additional evidence for the effectiveness of integrating visual and tactile stimuli in human–robot interaction.

Note that our aim is examining the effectiveness of touch stimuli for motivation enhancement, not whether they are the sole factor for improvement. In the context of applications for social robots or virtual agents designed to improve motivation, various interaction modalities are typically employed. Using touch stimuli (beyond visual touch) might provide a potential modality for interaction based on our experimental results, demonstrating the value of tactile sensation for motivation improvement.

Our study also suffers from several limitations. We investigated the effects of touch interaction using a specific virtual agent and a tactile stimulus. Although an avatar appearance is not the primary focus of this study, exploring such effects is critical for using different agents [30]. Related to this topic, another future work will investigate touch effects toward motivation improvements using different kinds of robots [31,32]. We also only visualized the participants' hands in the virtual environments; displaying the entire body or different appearances might yield different outcomes, similar to virtual agents' appearance.

16.7 CONCLUSION

Although touch interaction offers numerous positive effects in both human–robot interaction in physical environments and human–agent interaction in virtual environments, the impact of touch stimuli has not been extensively studied. We experimentally investigated the effects of touch stimuli by comparing the visual-only-touch and visual-tactile-touch effectiveness by a virtual agent in a virtual environment concerning motivation improvement effects. We developed a system that combines a virtual reality application and a physical robot to deliver both stimuli. Participants in the visual-tactile-touch condition performed more repetitive tasks than those in the visual-only-touch condition. These findings emphasize the importance of tactile stimuli in touch interaction for motivation improvement.

ACKNOWLEDGMENT

This work was supported by JST CREST Grant Number JPMJCR18A1, Japan.

REFERENCES

[1] J. B. F. van Erp, and A. Toet, "Social touch in human-computer interaction," *Frontiers in Digital Humanities*, vol. 2, no. 2, p. 2, 2015.

[2] M. Shiomi, H. Sumioka, and H. Ishiguro, "Survey of social touch interaction between humans and robots," *Journal of Robotics and Mechatronics*, vol. 32, no. 1, pp. 128–135, 2020.

[3] M. Shiomi, K. Nakagawa, K. Shinozawa, R. Matsumura, H. Ishiguro, and N. Hagita, "Does a robot's touch encourage human effort?," *International Journal of Social Robotics*, vol. 9, pp. 5–15, 2016.

[4] C. Bevan, and D. Stanton Fraser, "Shaking hands and cooperation in tele-present human-robot negotiation," In *Proceedings of the Tenth Annual ACM/IEEE International Conference on Human-Robot Interaction*, pp. 247–254, 2015.

[5] M. Shiomi, A. Nakata, M. Kanbara, and N. Hagita, "A hug from a robot encourages prosocial behavior," In *2017 26th IEEE International Symposium on Robot and Human Interactive Communication (RO-MAN)*, pp. 418–423, 2017.

[6] M. Shiomi, A. Nakata, M. Kanbara, and N. Hagita, "Robot reciprocation of hugs increases both interacting times and self-disclosures," *International Journal of Social Robotics*, vol. 13, pp. 353–361, 2020.

[7] N. Geva, F. Uzefovsky, and S. Levy-Tzedek, "Touching the social robot PARO reduces pain perception and salivary oxytocin levels," *Scientific Reports*, vol. 10, no. 1, p. 9814, 2020.

[8] M. Shiomi, and N. Hagita, "Audio-visual stimuli change not only robot's hug impressions but also its stress-buffering effects," *International Journal of Social Robotics*, vol. 13, pp. 469–476, 2021.

[9] X. Zheng, M. Shiomi, T. Minato, and H. Ishiguro, "What kinds of robot's touch will match expressed emotions?," *IEEE Robotics and Automation Letters*, vol. 5, pp. 127–134, 2019.

[10] X. Zheng, M. Shiomi, T. Minato, and H. Ishiguro, "How can robot make people feel intimacy through touch?," *Journal of Robotics and Mechatronics*, vol. 32, no. 1, pp. 51–58, 2020.

[11] X. Zheng, M. Shiomi, T. Minato, and H. Ishiguro, "Modeling the timing and duration of grip behavior to express emotions for a social robot," *IEEE Robotics and Automation Letters*, vol. 6, no. 1, pp. 159–166, 2020.

[12] Y. Okada, M. Kimoto, T. Iio, K. Shimohara, H. Nittono, and M. Shiomi, "Kawaii emotions in presentations: Viewing a physical touch affects perception of affiliative feelings of others toward an object," *PLoS One*, vol. 17, no. 3, p. e0264736, 2022.

[13] C. J. Willemse, and J. B. van Erp, "Social touch in human-robot interaction: Robot-initiated touches can induce positive responses without extensive prior bonding," *International Journal of Social Robotics*, vol. 11, no. 2, pp. 285–304, 2019.

[14] M. Teyssier, G. Bailly, C. Pelachaud, and E. Lecolinet, "Conveying emotions through device-initiated touch," *IEEE Transactions on Affective Computing*, vol. 13, pp. 1477–1488, 2020.

[15] J. Swidrak, and G. Pochwatko, "Being touched by a virtual human: Relationships between heart rate, gender, social status, and compliance," In *Proceedings of the 19th ACM International Conference on Intelligent Virtual Agents*, pp. 49–55, 2019.

[16] M. Botvinick, and J. Cohen, "Rubber hands 'feel' touch that eyes see," *Nature*, vol. 391, no. 6669, pp. 756–756, 1998.

[17] M. Tsakiris, and P. Haggard, "The rubber hand illusion revisited: Visuotactile integration and self-attribution," *Journal of Experimental Psychology: Human Perception and Performance*, vol. 31, no. 1, p. 80, 2005.

[18] M. Costantini, and P. Haggard, "The rubber hand illusion: Sensitivity and reference frame for body ownership," *Consciousness and Cognition*, vol. 16, no. 2, pp. 229–240, 2007.

[19] S. Shimada, K. Fukuda, and K. Hiraki, "Rubber hand illusion under delayed visual feedback," *PloS One*, vol. 4, no. 7, p. e6185, 2009.

[20] W. A. IJsselsteijn, Y. A. W. de Kort, and A. Haans, "Is this my hand I see before me? The rubber hand illusion in reality, virtual reality, and mixed reality," *Presence: Teleoperators and Virtual Environments*, vol. 15, no. 4, pp. 455–464, 2006.

[21] M. Slater, B. Spanlang, M. V. Sanchez-Vives, and O. Blanke, "First person experience of body transfer in virtual reality," *PLoS One*, vol. 5, no. 5, p. e10564, 2010.

[22] K. Higashino, M. Kimoto, T. Iio, K. Shimohara, and M. Shiomi, "Tactile stimulus is essential to increase motivation for touch interaction in virtual environment," *Advanced Robotics*, vol. 35, no. 17, pp. 1043–1053, 2021.

[23] F. Biocca, J. Kim, and Y. Choi, "Visual touch in virtual environments: An exploratory study of presence, multimodal interfaces, and cross-modal sensory illusions," *Presence: Teleoperators and Virtual Environments*, vol. 10, no. 3, pp. 247–265, 2001.

[24] M. Fusaro, M. Lisi, G. Tieri, and S. M. Aglioti, "Touched by vision: How heterosexual, gay, and lesbian people react to the view of their avatar being caressed on taboo body parts," 2020.

[25] K. Nagamachi, Y. Kato, M. Sugimoto, M. Inami, and M. Kitazaki, "Pseudo physical contact and communication in VRChat: A study with survey method in Japanese users," 2020.

[26] B. K. Jakubiak, and B. C. Feeney, "Keep in touch: The effects of imagined touch support on stress and exploration," *Journal of Experimental Social Psychology*, vol. 65, pp. 59–67, 2016.

[27] G. K. Essick, A. James, and F. P. McGlone, "Psychophysical assessment of the affective components of non-painful touch," *Neuroreport*, vol. 10, no. 10, pp. 2083–2087, 1999.

[28] K. Higashino, M. Kimoto, T. Iio, K. Shimohara, and M. Shiomi, "Effects of social touch from an agent in virtual space: Comparing visual stimuli and virtual-tactile stimuli," In *2020 29th IEEE International Conference on Robot and Human Interactive Communication (RO-MAN)*, pp. 768–774, 2020.

[29] C. Bartneck, D. Kulić, E. Croft, and S. Zoghbi, "Measurement instruments for the anthropomorphism, animacy, likeability, perceived intelligence, and perceived safety of robots," *International Journal of Social Robotics*, vol. 1, no. 1, pp. 71–81, 2009.

[30] M. Kimoto, Y. Otsuka, M. Imai, and M. Shiomi, "Effects of appearance and gender on pre-touch proxemics in virtual reality," *Frontiers in Psychology*, vol. 14, 2023.

[31] D. F. Glas, T. Minato, C. T. Ishi, T. Kawahara, and H. Ishiguro, "Erica: The erato intelligent conversational android," In *2016 25th IEEE International Symposium on Robot and Human Interactive Communication (RO-MAN)*, pp. 22–29, 2016.

[32] M. Shiomi, H. Sumioka, K. Sakai, T. Funayama, and T. Minato, "SŌTO: An android platform with a masculine appearance for social touch interaction," In *Companion of the 2020 ACM/IEEE International Conference on Human-Robot Interaction*, pp. 447–449, 2020.

CHAPTER 17

Understanding Self-Touch Behaviors and Stress-Buffering Effects

Ayumi Hayashi, Emi Anzai, Naoki Saiwaki, Hidenobu Sumioka, and Masahiro Shiomi

Advanced Telecommunications Research Institute International, Kyoto, Japan

Nara Women's University, Nara, Japan

17.1 INTRODUCTION

Touch interactions with close individuals offer numerous benefits, such as decreased heart rates and blood pressure [1, 2], enhanced immune systems [3], stimulation of early development [4], emotional communication [5], and positive emotions [6, 7]. Even imagined touch interactions provide stress-buffering effects [8], and similar benefits can be observed in touch interactions with social robots [9].

Unfortunately, the COVID-19 pandemic led to the imposition of physical barriers on interaction with others, leading to touch starvation, a growing social issue that elevates stress levels because touch is a vital form of human communication [10].

DOI: 10.1201/9781003384274-22

FIGURE 17.1   Participant self-touch during a stressful situation.

Therefore, our focus shifted to a more straightforward touch interaction: intra-active or self-touch [11]. Although self-touch has been extensively studied in human science literature [11] (Section 17.2), most research has focused on its positive effects and occurrences. Our study promotes self-touch and actively leverages its positive effects through a wearable system.

In this study, we proposed a wearable system that identifies users' self-touches and provides supportive audio feedback to reinforce their mental states and promote additional self-touches. By employing a fabric-based touch sensor and incorporating a stressful task in our experimental design (Figure 17.1), we assessed our developed system to address the following questions:

1. Do self-touches increase due to supportive voices triggered by self-touch?

2. Do increased self-touch behaviors and supportive voices enhance stress-buffering effects?

Note that this chapter is modified based on our previous work [12], edited to be comprehensive and fit with the context of this book.

17.2 RELATED WORKS

Various studies have focused on the impact of self-touches, including perceived pain and behavior changes. For example, past research works have shown that a self-touch can reduce the intensity of perceived pain during a thermal grill illusion [13] and improve focus and attitude extremity toward a specific target, regardless of conscious or unconscious awareness [14]. Self-touch can also influence motor imagery, such as reaction time, signifying behavior changes in those engaging in it [15].

Researchers have also investigated the circumstances surrounding self-touch behaviors, such as how and when people touch themselves. Past studies found associations between self-touch patterns on the face and perceived cognitive and emotional load [16]. Video-based surveys revealed that self-touches can display engagement [11]. In terms of social perception, observers rated colleagues who self-touched more positively [15], and another research indicated that self-touch behaviors create warm and expressive impressions based on which body parts are touched [16].

Furthermore, although the influence of gender and its roles (interviewers or applicants) on self-touch behaviors has been investigated in stressful situations like job interviews [17], the relationship between perceived stress and self-touch frequency remains unclear. A past study reported an increase in self-touch behaviors due to stress [18], while another study described contrary findings [19]. Another work reported that self-touch is positively correlated with people's state and anxiety or negatively correlated with agreeableness [20].

Even though previous research identified the positive effects of self-touch, there has been insufficient encouragement of self-touch behaviors and its benefits. Our work is distinguished by two unique aspects: (1) the development of a system that detects users' self-touches and provides supportive voices and (2) an experimental evaluation of our system's effectiveness with human participants.

17.3 EXPERIMENT DESIGN AND SYSTEM OVERVIEW

17.3.1 Experiment Task

In our experiment, we used the Trier Social Stress Test (TSST) [21]. Initially, participants rested for five minutes, followed by a five-minute preparation period for a three-minute, dummy job interview presentation. After the presentation, a debriefing session was conducted by the experimenter.

Because this experiment was conducted under COVID-19 pandemic restrictions, we carried out the TSST sessions in our research facilities by Zoom. Participants sat alone in a room to simulate an online job interview. To provide consistent audio stimuli for the TSST, we used pre-recorded voices and a voice changer to create a masculine-sounding voice from one of the authors' recorded voices. The experimenter managed the timing of the interviewer's voice inputs.

17.3.2 Experiment System

17.3.2.1 Touch Sensor

The touch sensor used in this experiment is a fabric-based device that measures the capacitance changes to detect a human's physical contact, i.e., self-touches. With a measurement frequency of approximately 10 Hz, the system identifies touch/no-touch based on a pre-determined threshold value.

17.3.2.2 Supportive and Interviewer Voice Control System

This system manages two types of voices: supportive and interviewer. Speech synthesis software generated the content of two feminine voices, which were delivered through earphones in response to the touch sensor outputs.

We prepared 13 voices for the before-interviews, which included such advice for preparing and giving a presentation as narrative flow, presentation speed, and mental support that encouraged relaxation. For the during-interviews, we prepared five voices that gave simple responses, e.g., "uh-huh," to avoid interrupting the presentation. For the after-interviews, we prepared ten voices that offered encouragement and presentation feedback. The supportive voice types were autonomously changed due to the interview phases. When the participants touched the sensor, the system randomly played one of the voices to avoid repetition. We prepared eight interviewer voices for the TSST task, including asking the participants to continue if they failed to use the entire five minutes allotted for their presentation.

During the interview, we used five simple response voices (e.g., uh-huh) to minimize interruptions. For the after-interview phase, ten voices provided encouragement and presentation feedback. The system autonomously adjusted to the appropriate supportive voice types based on the interview phase and randomly chose voice files to ensure a variety of responses.

17.3.2.3 Operator GUI for TSST Task Management

Our GUI contains several buttons that allow the experimenter to manage the audio stimuli for the TSST task. The experimenter can control the interviewer voices and change the supportive voice types throughout different interview phases.

17.4 EXPERIMENT

17.4.1 Hypotheses and Predictions

The advantages of self-touch behaviors have been scrutinized [22] and extended to both conscious and unconscious styles of self-touch [14]. Since we believe that self-touch can be a stress buffer, we investigated its effectiveness for stress reduction by designing a system whose supportive voices encourage conscious self-touch; these voices are activated based on a touch sensor. If our system works successful, we believe that it will increase self-touch frequency and subsequently enhance stress-buffering effects. Based on these contexts, we made the following predictions:

1. Participants using our system will engage in self-touch behaviors more frequently than those who do not use it.

2. Participants using our system will experience reduced stress levels than those who do not use it.

17.4.2 Conditions

Our experiment followed a between-participants design and randomly assigned participants to one of two groups:

– **Proposed condition:** In this condition, our system is used by the participants during the experiment. Thus, the system provides supportive voices based on their own self-touches. Eight men and five women participated in this group.

– **Alternative condition:** In this condition, during the experiment, the participants did not use our system, although they wore a touch sensor. Thus, the system did not provide supportive voices even though self-touches of the participants were detected. Ten men and seven women participated in this group.

The imbalanced number of participants and gender ratios reflect the COVID-19 state of emergency, sometimes leading to last minute cancellations.

In our analysis, we considered the gender factor since previous research has indicated differences in self-touch behavior patterns based on gender [17]. Several touch interaction studies have also reported gender effects, even when self-touch was not the primary focus [23]. We incorporated the gender factor into the between-participants design.

17.4.3 Environment

We installed a display and a speaker in a room of our laboratory to simulate an online job interview by Zoom. The experimenter in another room controlled the timing of the audio stimuli for the interviews. Supportive system voices were delivered through earphones. The speaker and earphone volumes were adjusted to ensure that participants could simultaneously hear both the audio stimuli from Zoom and our system.

17.4.4 Participants

Thirty participants took part in the experiment: 17 women and 13 men. Their ages ranged from 21 to 55, with an average of 38.1.

17.4.5 Measurements

We assessed the stress-buffering effects by measuring the salivary amylase levels, which typically rise when individuals are experiencing stress. We recorded the difference in salivary amylase levels before and after the experiment; a lower value signified enhanced stress-buffering effects.

Moreover, we counted the number of participants' self-touch behaviors during the experiment. Given the challenge of determining whether a self-touch is conscious or unconscious, we simply counted the number of self-touch behaviors at the sensor location. To examine the impact of the TSST task, we categorized the number of self-touch behaviors into three distinct phases corresponding to the supportive voice settings: before-interview, during-interview, and after-interview.

17.4.6 Procedure

The experimenter provided a detailed explanation to the participants, who then gave written informed consent to join. Our institute's ethics committee approved the experimental procedure. Before and after the TSST task, we measured the salivary amylase levels. At the experiment's end, the experimenter conducted a debriefing session.

17.5 RESULTS

17.5.1 Verification of Prediction 1

Figure 17.2 shows the average and S.E. values of the number of self-touch behaviors. We conducted a three-factor mixed ANOVA for each scale on phase, condition, and gender and identified significant main effects in the phase factor ($F (2, 52) = 3.606$, $p = 0.034$, partial $\eta^2 = 0.122$) and in the condition factor ($F (1, 36) = 5.672$, $p = 0.025$, partial $\eta^2 = 0.179$). We found

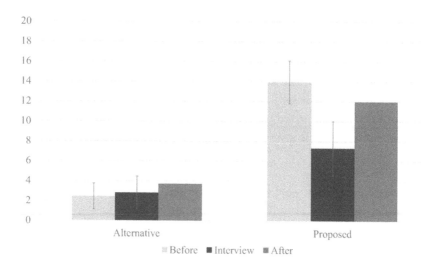

FIGURE 17.2 Number of self-touch behaviors (significant difference appears in condition factor, regardless of phase factor).

significant trends in the simple interaction effect between condition and gender (F (1, 26) = 3.066, p = 0.092, partial η^2 = 0.105) and a simple interaction effect between phase and condition (F (2, 52) = 2.989, p = 0.059, partial η^2 = 0.103). No significance was found in the gender factor (F (1, 26) = 1.484, p = 0.234, partial η^2 = 0.054), in the simple interaction effect between phase and gender (F (2, 52) = 1.690, p = 0.194, partial η^2 = 0.061), or in the two-way interaction effect (F (2, 52) = 1.049, p = 0.357, partial η^2 = 0.039).

Thus, participants who used our system engaged in self-touch behaviors more frequently than those who did not; prediction 1 was supported.

17.5.2 Verification of Prediction 2

Figure 17.3 shows the average and SE values of the differences in salivary amylases. We conducted a two-factor mixed ANOVA for each scale on condition and gender and identified significant main effects in the interaction effect (F (1, 26) = 4.408, p = 0.046, partial η^2 = 0.145). No significance was found in the condition factor (F (1, 26) = 1.191, p = 0.285, partial η^2 = 0.044) or in the gender factor (F (1, 26) = 1.308, p = 0.263, partial η^2 = 0.048). Multiple comparisons with the Bonferroni method revealed significant differences in the simple main effects of the condition in men (alternative > proposed (p = 0.045)). Other combinations did not reveal any significant differences.

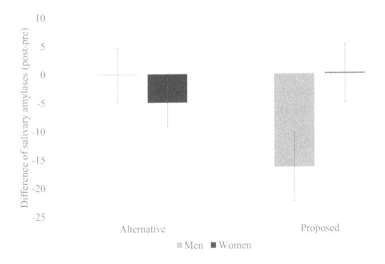

FIGURE 17.3　The average and S.E. values of the differences in salivary amylases.

Thus, men who used our system experienced reduced stress levels compared to those who did not, whereas women who used our system did not experience reduced stress levels compared to those who did not; prediction 2 was partially supported.

17.6 DISCUSSION

Our results demonstrated that the developed system increased the frequency of self-touch behaviors; however, stress-buffering effects were only evident among men in our experiment. One possible explanation is that men engaged in more self-touch behaviors than women, which may have influenced their perceived stress. A potential contributing factor to these findings could be the gender of the supportive voice, which was generated using feminine-speech synthesis software. To understand the influence of voice gender in the context of self-touch related voices, additional experiments must be performed.

　In this study, we concentrated on the effectiveness of self-touch behaviors. To our knowledge, no research has directly compared stress-buffering effects among different touch interactions, e.g., comparisons between self-touch and non-self-touches. Previous studies investigated the effects of non-self-touches with intimate persons [4], imagined individuals [8], animals/pets [24], virtual agents [25], and social robots [26]. Therefore, another possible future work might compare the effects of stress-buffering effects with these different entities. In comparisons between self-touches and touches by agents including robots, their appearances are one essential factor; for example, using android robots with human-like appearances [27–29], robot-like appearances robots [30–32], and pet-like appearance robots [33–36] might change the touch effects. Another possible factor is the touch characteristics themselves because expressing emotions by touch may have influenced the stress-buffering effects [37–40].

17.7 CONCLUSION

We focused on the effects of encouraging self-touch behaviors on stress-buffering effects and developed a system that integrates a fabric-based touch sensor with a supportive voice feature to promote self-touch activities. To investigate our developed system's effectiveness, we conducted an experiment in which participants performed a stressful task (TSST) and evaluated the stress-buffering effect of the developed system. Our experiment results showed that the developed system significantly increased the number of self-touch behaviors under stressful tasks. Unfortunately, the stress-buffering effects were only found in men.

ACKNOWLEDGMENT

This research work was supported in part by JST CREST Grant Number JPMJCR18A1, Japan (sensor development), and JST, Moonshot R&D Grant Number JPMJMS2011 (experiment and analysis).

REFERENCES

[1] K. C. Light, K. M. Grewen, and J. A. Amico, "More frequent partner hugs and higher oxytocin levels are linked to lower blood pressure and heart rate in premenopausal women," *Biological Psychology*, vol. 69, no. 1, pp. 5–21, 2005.

[2] K. M. Grewen, B. J. Anderson, S. S. Girdler, and K. C. Light, "Warm partner contact is related to lower cardiovascular reactivity," *Behavioral Medicine*, vol. 29, no. 3, pp. 123–130, 2003.

[3] S. Cohen, D. Janicki-Deverts, R. B. Turner, and W. J. Doyle, "Does hugging provide stress-buffering social support? A study of susceptibility to upper respiratory infection and illness," *Psychological Science*, vol. 26, no. 2, pp. 135–147, 2015.

[4] T. Field, "Touch for socioemotional and physical well-being: A review," *Developmental Review*, vol. 30, no. 4, pp. 367–383, 2010.

[5] M. J. Hertenstein, R. Holmes, M. McCullough, and D. Keltner, "The communication of emotion via touch," *Emotion*, vol. 9, no. 4, p. 566, 2009.

[6] J. K. Burgoon, D. B. Buller, J. L. Hale, and M. A. Turck, "Relational messages associated with nonverbal behaviors," *Human Communication Research*, vol. 10, no. 3, pp. 351–378, 1984.

[7] K. Takemura, "The effect of interpersonal sentiments on behavioral intention of helping behavior among Japanese students," *The Journal of Social Psychology*, vol. 133, no. 5, pp. 675–681, 1993.

[8] B. K. Jakubiak, and B. C. Feeney, "Keep in touch: The effects of imagined touch support on stress and exploration," *Journal of Experimental Social Psychology*, vol. 65, pp. 59–67, 2016.

[9] M. Shiomi, A. Nakata, M. Kanbara, and N. Hagita, "Robot reciprocation of hugs increases both interacting times and self-disclosures," *International Journal of Social Robotics*, vol. 13, pp. 353–361, 2021.

[10] J. Durkin, D. Jackson, and K. Usher, "Touch in times of COVID-19: Touch hunger hurts," *Journal of Clinical Nursing*, vol. 30, pp. e4–e5, 2021.

[11] S. J. Bolanowski, R. T. Verrillo, and F. McGlone, "Passive, active and intra-active (self) touch," *Somatosensory & Motor Research*, vol. 16, no. 4, pp. 304–311, 1999.

[12] A. Hayashi, E. Anzai, N. Saiwaki, H. Sumioka, and M. Shiomi, "Does encouraging self-touching behaviors with supportive voices increase stress-buffering effects?," In *Proceedings of the 2022 ACM/IEEE International Conference on Human-Robot Interaction*, Sapporo, Hokkaido, Japan, pp. 787–791, 2022.

[13] M. P. Kammers, F. De Vignemont, and P. Haggard, "Cooling the thermal grill illusion through self-touch," *Current Biology*, vol. 20, no. 20, pp. 1819–1822, 2010.

[14] A. Kronrod, and J. M. Ackerman, "I'm so touched! Self-touch increases attitude extremity via self-focused attention," *Acta Psychologica*, vol. 195, pp. 12–21, 2019.

[15] M. Conson, E. Mazzarella, and L. Trojano, "Self-touch affects motor imagery: A study on posture interference effect," *Experimental Brain Research*, vol. 215, no. 2, p. 115, 2011.

[16] S. M. Mueller, S. Martin, and M. Grunwald, "Self-touch: Contact durations and point of touch of spontaneous facial self-touches differ depending on cognitive and emotional load," *PLoS One*, vol. 14, no. 3, p. e0213677, 2019.

[17] S. Goldberg, and R. Rosenthal, "Self-touching behavior in the job interview: Antecedents and consequences," *Journal of Nonverbal Behavior*, vol. 10, no. 1, pp. 65–80, 1986.

[18] N. D. Butzen, V. Bissonnette, and D. McBrayer, "Effects of modeling and topic stimulus on self-referent touching," *Perceptual and Motor Skills*, vol. 101, no. 2, pp. 413–420, 2005.

[19] R. Ackerley, E. Hassan, A. Curran, J. Wessberg, H. Olausson, and F. McGlone, "An fMRI study on cortical responses during active self-touch and passive touch from others," *Frontiers in Behavioral Neuroscience*, vol. 6, p. 51, 2012.

[20] H. T. Pang, F. Canarslan, and M. Chu, "Individual differences in conversational self-touch frequency correlate with state anxiety," *Journal of Nonverbal Behavior*, vol. 46, no. 3, pp. 299–319, 2022.

[21] C. Kirschbaum, K.-M. Pirke, and D. H. Hellhammer, "The 'Trier Social Stress Test'—A tool for investigating psychobiological stress responses in a laboratory setting," *Neuropsychobiology*, vol. 28, no. 1–2, pp. 76–81, 1993.

[22] K. Densing, H. Konstantinidis, and M. Seiler, "Effect of stress level on different forms of self-touch in pre- and postadolescent girls," *Journal of Motor Behavior*, vol. 50, no. 5, pp. 475–485, 2018.

[23] J. D. Fisher, M. Rytting, and R. Heslin, "Hands touching hands: Affective and evaluative effects of an interpersonal touch," *Sociometry*, vol. 39, no. 4, pp. 416–421, 1976.

[24] N. Guéguen, "Touch, awareness of touch, and compliance with a request," *Perceptual and Motor Skills*, vol. 95, no. 2, pp. 355–360, 2002.

[25] P. Sykownik, and M. Masuch, "The experience of social touch in multi-user virtual reality," In *26th ACM Symposium on Virtual Reality Software and Technology*, Ottawa, Canada, pp. 1–11, 2020.

[26] M. Shiomi, H. Sumioka, and H. Ishiguro, "Survey of social touch interaction between humans and robots," *Journal of Robotics and Mechatronics*, vol. 32, no. 1, pp. 128–135, 2020.

[27] D. F. Glas, T. Minato, C. T. Ishi, T. Kawahara, and H. Ishiguro, "Erica: The erato intelligent conversational android," In *2016 25th IEEE International Symposium on Robot and Human Interactive Communication (RO-MAN)*, New York, NY, United States, pp. 22–29, 2016.

[28] D. Sakamoto, and H. Ishiguro, "Geminoid: Remote-controlled android system for studying human presence," *Kansei Engineering International*, vol. 8, no. 1, pp. 3–9, 2009.

[29] M. Shiomi, H. Sumioka, K. Sakai, T. Funayama, and T. Minato, "SŌTO: An android platform with a masculine appearance for social touch interaction," In *Companion of the 2020 ACM/IEEE International Conference on Human-Robot Interaction*, Cambridge, United Kingdom, pp. 447–449, 2020.

[30] T. Kanda, H. Ishiguro, T. Ono, M. Imai, and R. Nakatsu, "Development and evaluation of an interactive humanoid robot "Robovie"," In *Proceedings 2002 IEEE International Conference on Robotics and Automation (Cat. No.02CH37292)*, Washington, DC, United States, pp. 1848–1855 vol.2, 2002.

[31] R. Matsumura, M. Shiomi, K. Nakagawa, K. Shinozawa, and T. Miyashita, "A desktop-sized communication robot: "Robovie-mr2"," *Journal of Robotics and Mechatronics*, vol. 28, no. 1, pp. 107–108, 2016.

[32] R. Matsumura, and M. Shiomi, "An animation character robot that increases sales," *Applied Sciences*, vol. 12, no. 3, p. 1724, 2022.

[33] T. Shibata, "An overview of human interactive robots for psychological enrichment," *Proceedings of the IEEE*, vol. 92, no. 11, pp. 1749–1758, 2004.

[34] M. Fujita, "AIBO: Toward the era of digital creatures," *The International Journal of Robotics Research*, vol. 20, no. 10, pp. 781–794, 2001.

[35] N. Yoshida, S. Yonemura, M. Emoto, K. Kawai, N. Numaguchi, H. Nakazato, S. Otsubo, M. Takada, and K. Hayashi, "Production of character animation in a home robot: A case study of lovot," *International Journal of Social Robotics*, vol. 14, no. 1, pp. 39–54, 2022.

[36] K. Nakagawa, R. Matsumura, and M. Shiomi, "Effect of robot's play-biting in non-verbal communication," *Journal of Robotics and Mechatronics*, vol. 32, no. 1, pp. 86–96, 2020.

[37] X. Zheng, M. Shiomi, T. Minato, and H. Ishiguro, "What kinds of robot's touch will match expressed emotions?," *IEEE Robotics and Automation Letters*, vol. 5, pp. 127–134, 2019.

[38] X. Zheng, M. Shiomi, T. Minato, and H. Ishiguro, "How can robot make people feel intimacy through touch?," *Journal of Robotics and Mechatronics*, vol. 32, no. 1, pp. 51–58, 2019.

[39] X. Zheng, M. Shiomi, T. Minato, and H. Ishiguro, "Modeling the timing and duration of grip behavior to express emotions for a social robot," *IEEE Robotics and Automation Letters*, vol. 6, no. 1, pp. 159–166, 2020.

[40] M. Shiomi, X. Zheng, T. Minato, and H. Ishiguro, "Implementation and evaluation of a grip behavior model to express emotions for an android robot," *Frontiers in Robotics and AI*, vol. 8, p. 755150, 2021.

CHAPTER 18

Mediated Hug Modulates Impressions of Hearsay Information

Junya Nakanishi, Hidenobu Sumioka, and Hiroshi Ishiguro

Osaka University, Osaka, Japan

Advanced Telecommunications Research Institute International, Kyoto, Japan

18.1 INTRODUCTION

Tele-operated humanoid robots serve as virtual extensions of individuals in remote communication scenarios. These humanoid robots, remotely controlled by an operator, can convey the operator's voice and non-verbal expressions to a remote person. This simulation of face-to-face interactions, termed "telepresence" [1, 2], significantly enriches the remote communication experience. A distinct advantage of robot-mediated communication over traditional telecommunication methods (such as audio or video) is physical interaction. Touching a tele-operated humanoid robot can create the illusion of physical contact with the operator. Recent findings suggest that such mediated interpersonal touch can yield similar positive effects to those of direct interpersonal touch [3, 4].

Interpersonal touch (e.g., a handshake, a pat on the back, a kiss, or a hug) provides significant influence in various interaction scenarios [5]. Touching acts as a psychosomatic stabilizer, a messenger, or an

DOI: 10.1201/9781003384274-23

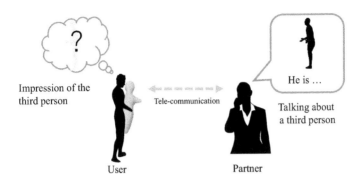

FIGURE 18.1 Impression of hearsay information about an absentee third person provided by a conversation.

attitude modulator [6] notably contribute to human development and social communication [7, 8]. Interpersonal touch also plays a pivotal role in shaping impressions during human communication. Research indicates that recipients of touching form a more favorable impression of the touch provider compared to non-touch conditions [9–13]. Moreover, this positive impression extends beyond the touch provider to encompass information associated with them, such as their organization [9, 13]. Given that positive impressions can lead to positive behavior [14–16], the impression bias generated by interpersonal touch has garnered considerable interest.

Our study explores the effect of mediated hugs on the impression formation of hearsay information, specifically the contents of a partner's conversation. To examine this impression bias, we conducted an experiment in which participants evaluated by a questionnaire and a recall test the information provided about a third person's behavior. We compared the effects of a mobile speaker with a huggable communication medium called "Hugvie," which produces mediated hugs. Our findings prompt a discussion about a model and how mediated hugs influence impression formations (Figure 18.1).

18.2 BACKGROUND

18.2.1 Implementation of Mediated Interpersonal Touch

Advanced telecommunication technologies have facilitated the concept of "virtual interpersonal touch" or "mediated social touch" between remote individuals [3, 4, 17]. This idea can be implemented in two main ways. One approach involves recreating tactile stimulation based on input from a remote partner through a wearable device [18–21]. For instance, Cabibihan et al. demonstrated that a mediated hand touch on an arm reduced participants' heart rates after they watched a sad video clip [18].

The second implementation involves human-like physical embodiments such as telecommunication media, and these embodiments serve as avatars for the operators [1,22,23]. For example, tele-operated humanoid robots can convey an operator's touch, physical motion, voice, and attitude, conveying to users the impression of directly interacting with an operator [2]. Our study focuses on this latter form of mediated interpersonal touch.

18.2.2 Mediated Hug

Various mediated hug devices have been developed [24–31]. For example, DiSalvo et al.'s "The Hug" mimics the shape and gesture of a human hug to explore intimate communication across distances [32]. Sumioka et al. reported on the stress reduction effect of a mediated hug produced by Hugvie, a huggable communication medium [33].

Previous studies have described the effect of mediated hugs on impressions of communication partners such as likeability [34], interest [35], and trust [36]. However, unclear domains remain in impression bias caused not only by mediated hugs but also by other types of mediated interpersonal touch, especially regarding hearsay information. Understanding impression bias in hearsay information is becoming increasingly important for successful human communication in the field of social psychology [37–39].

18.2.3 Hypothesis

Both direct and mediated interpersonal touch have positively influenced the impressions of communication partners [7–11,35–37]. Moreover, direct interpersonal touch enhances the impression of information related to a communication partner [7,11]. As such, we hypothesize that a mediated hug could potentially lead to a more favorable impression of hearsay information.

18.3 EXPERIMENT

We performed a laboratory study that incorporated both subjective and objective assessment methods and specifically utilized a survey and a recall test to substantiate our hypothesis. The survey gauged both the overall and specific perceptions of the transmitted information. The recall test assessed the comprehensive impressions of it in terms of memory and interpretation. We compared a standard Bluetooth speaker and a huggable communication device called Hugvie. This study was approved by the Ethics Committee of the Advanced Telecommunications Research Institute International (Kyoto, Japan).

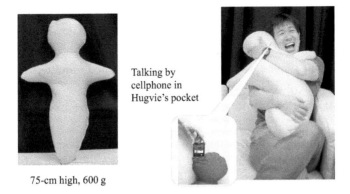

Talking by
cellphone in
Hugvie's pocket

75-cm high, 600 g

FIGURE 18.2   Hugvie: a huggable communication medium.

18.3.1 Apparatus

Hugvie (Figure 18.2) is designed to provide a hug during communication. It is a soft, human-shaped cushion filled with polystyrene microbeads and enclosed in a mixed-fiber covering of acrylic and rayon. By inserting a hands-free mobile device or speaker into a pocket in the cushion's "head," users can converse while embracing it. This combination of physically embracing the device and hearing a partner's voice emanating from a location near the user's ear fosters a sense of the partner's proximity and strengthens a mutual bond. We transmitted auditory cues from the conversation partner with a portable Bluetooth speaker (MOT-EQ5, MOTOROLA). The same speaker was also employed in the standard speaker condition to maintain consistency in audio stimuli properties.

18.3.2 Experimental Design and Procedure

Our study employed a between-subjects design to compare responses to a Bluetooth speaker and Hugvie. Our participants were young Japanese individuals (average age = 22.4 years, SD = 3.57, 18 males and 10 females). Each condition incorporated nine males and five females. All participants were briefed on the study and provided informed consent.

First, the participants in the Hugvie group were shown how to use the huggable robot and the speaker group was introduced to the Bluetooth speaker. As a trial to familiarize them with listening to the story using their assigned device, participants then listened to a 12-minute excerpt from "The Fall of Freddie the Leaf" by Leo Buscaglia by the device. We used a pre-recorded female voice for this phase.

Participants were then instructed to listen to an explanation about a third male individual they did not know, which was communicated to them by their assigned device. A two-minute audio recording of this description was played using the same pre-recorded female voice. After listening, participants evaluated their impressions on the third person by a questionnaire (details provided below in the "measurements" section).

Subsequently, participants engaged in a ten-minute fake task in which they viewed a conversational story by a text chat application on a screen and evaluated the experience through a second questionnaire. This fake task, as suggested in previous studies [37–39], occupied the participants' temporary memory, allowing us to investigate the impressions preserved in their long-term memory, which exerts much influence over the everyday impressions formed of others. Participants then undertook a recall test and wrote down what they remembered about the third individual based on the explanation they heard.

We used the ambiguous passages created by Echterhoff et al. for the content of the third individual's information [39]. They consisted of six paragraphs, each containing two passages that imply slightly contrasting traits. The following are the 12 traits: moral/self-righteous, cultivated/artificial, adventurous/reckless, independent/aloof, persistent/stubborn, and thrifty/stingy. For instance, the "thrifty or stingy" trait was depicted with the sentence: "To improve his life, he tries to save money. He uses coupons, buys things on sale, and avoids donating money to charity or lending money to friends." This design allowed us to easily observe any bias in the participants' impressions of the third individual. To reduce bias that might arise from the participants' individual perceptions of names, we altered the original "Michael" to a neutral "he." The passages were translated into Japanese for the convenience of the participants.

18.3.3 Measurements and Analysis

We evaluated the participants' impressions of the third person by a questionnaire and a recall test. The collected data were analyzed using either a Welch Two Sample t-test or a Mann-Whitney U test, depending on the data's normality. The Shapiro-Wilk normality test determined the appropriate statistical test. In cases where the data were normally distributed ($p < 0.05$), a Welch Two Sample t-test was applied; for non-normally distributed data, a Mann-Whitney U test was utilized.

18.3.4 Questionnaire

To assess the overall impression of the third person, participants rated their personal preferences on an 11-point Likert scale, ranging from −5 (dislike) to +5 (like). Furthermore, their impressions of the third person's traits were gauged using an 11-point Likert scale. These traits corresponded to six pairs of slightly contrasting traits, mirroring those found in the passages: moral versus self-righteous, cultivated versus artificial, adventurous versus reckless, independent versus aloof, persistent versus stubborn, and thrifty versus stingy. The average scores for each item were then compared between the Hugvie and speaker conditions.

18.3.5 Recall Test

Participants were given three minutes to record their recollections of the information about the third person to which they previously listened. Positive and negative responses were tallied based on the original methodology [39]. However, some participants wrote down information that was not provided or imagined details based on the given information: "He often donates money to charity" or "He is a stubborn person." To handle these responses, two independent coders (one male, one female) who were unaware of the study's purpose categorized the responses into original and non-original information, with the original information corresponding to 12 original traits and everything else fell under non-original. The coders came to a mutual agreement after discussion.

The non-original information was then assessed by an additional group of fourteen participants using an 11-point Likert scale, where +5 represents positive and −5 represents negative. Using a t-test, we determined the positivity or negativity of each passage by comparing its score to zero ($p<0.05$). This yielded 12 positive, 11 negative, and 12 neutral responses. We then compared the number of positive and negative responses within the original and non-original information.

18.3.6 Correlation

To examine factors linked to personal preference, we also computed a correlation coefficient using a correlation test between personal preference and the other assessed items. If normality was confirmed by the Shapiro-Wilk normality test ($p<0.05$), we used a Pearson correlation test. In cases where it was not confirmed, we applied a Spearman's rank correlation test.

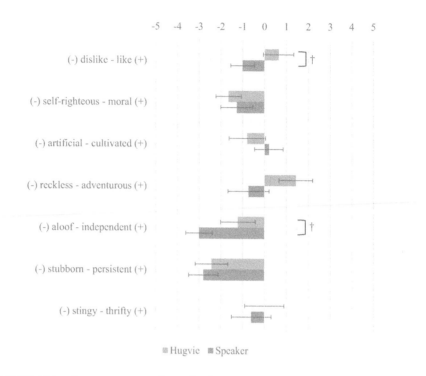

FIGURE 18.3   Average score and standard error in a questionnaire (†$p<0.1$).

18.4 RESULT

None of the questionnaire results indicated a significant difference between the Hugvie and speaker conditions (Figure 18.3): like–dislike of the third person ($t(24.66)=1.80$, $p=0.084$), moral–self-righteous ($t(24.77)=-0.51$, $p=0.61$), cultivated–artificial ($t(24.58)=-1.11$, $p=0.28$), adventurous–reckless ($U=128.5$, $p=0.16$), independent–aloof ($U=136.5$, $p=0.077$), persistent–stubborn ($U=107$, $p=0.69$), and thrifty–stingy ($t(25.99)=0.38$, $p=0.71$). However, like–dislike and the independent–aloof scale showed a marginal significance ($p<0.1$).

A recall test revealed a significant difference in the negative information recall between the Hugvie and speaker conditions (original: $U=87$, $p=0.61$; non-original: $U=53$, $p=0.024$), although there was no significant difference found in the positive information recall (original: $U=104.5$, $p=0.77$; non-original: $U=101.5$, $p=0.88$). Participants in the Hugvie condition recalled less non-original negative information about the target person than those in the speaker condition (Figure 18.4). Table 18.1 presents correlation coefficients between personal preferences to the third person (like–dislike scale) and other variables (positive original: $S=3,272.6$,

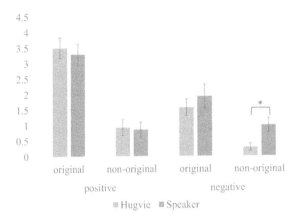

FIGURE 18.4   Average score and standard error in a recall test ($*p < 0.05$).

TABLE 18.1   Correlation Coefficients between Personal
Preference and Other Items ($\dagger p < 0.1$, $*p < 0.05$, $**p < 0.01$)

Item		Like–Dislike
Positive	Original	0.10
	Non-original	0.37†
Negative	Original	−0.14
	Non-original	−0.40*
Moral–self-righteous		0.15
Cultivated–artificial		0.48**
Adventurous–reckless		0.09
Independent–aloof		0.28
Persistent–stubborn		0.17
Thrifty–stingy		0.43*

$p=0.60$; positive non-original: $S=2,298$, $p=0.052$; negative original: $S=4,152.7$, $p=0.49$; negative non-original: $S=5,105.8$, $p=0.036$; moral-self-righteous: $t(26)=0.871$, $p=0.39$; cultivated–artificial: $t(26)=2.7953$, $p=0.0096$; adventurous–reckless: $S=3,317$, $p=0.64$; independent–aloof: $S=2,615.8$, $p=0.14$; persistent–stubborn: $S=3,026.1$, $p=0.38$; thrifty-stingy: $S=2,085.2$, $p=0.023$).

18.5 DISCUSSION

Our findings lend support to our hypothesis: mediated hugging can induce a more favorable impression of hearsay information. This finding is suggested by both the subjective and objective evaluations indicating that Hugvie users formed a more positive impression of the shared information. The likeability scale on the questionnaire revealed that participants using

Hugvie had a marginally more positive overall impression of the third person. Furthermore, the recall test indicated that participants in the Hugvie condition remembered fewer instances of non-original negative information about the third person. A significant correlation was also found between likeability and the recall of non-original negative information, demonstrating consistency between the subjective and objective evaluations.

Identifying the impression aspects that were enhanced sheds light on the mechanism behind the impression bias caused by mediated hugging. Regarding the questionnaire, a marginally significant difference was observed in the independent–aloof scale. While participants in both conditions did not view the third person as independent, those in the speaker condition perceived him as more aloof than the participants in the Hugvie condition. This could be attributed to the soft tactile stimulation provided by Hugvie, aligning with previous findings that such stimulation can induce less rigid or strict impressions of personality [40]. However, we did not find a significant correlation between likeability and the independent–aloof scale, suggesting that the latter may not be a key factor in the overall impression formation from hearsay information.

With respect to the recall test, we observed a significant difference in the recall of the non-original negative information between the Hugvie and speaker conditions. This finding illuminates the positive influence that mediated hugging can have on impression formation. In psychology, individuals with anxiety or depression are more likely to negatively interpret ambiguous social events and focus more on negative stimuli [41]. We speculate that these negative interpretations that focus on negativity can contribute to negative impression formation. Prior research has identified the stress-relieving effects of virtual hugging, such as reductions in stress hormones and stress-related emotions [33–35]. These studies suggest a possible mechanism in which mediated hugging reduces participants' stress, thereby lessening negative interpretations or attention to negative stimuli during impression formation. The observed reduction in the non-original negative information in our study supports this mechanism. Additionally, our model could be extended to other forms of direct and mediated interpersonal touch beyond hugging, given the established psychological and physiological effects of such touch on stress reduction [18,42–46].

Our study does, however, have its limitations. For instance, we only used a female voice as the audio stimulus, meaning our results might vary if a male voice were used. The gender of the third person, assumed to be male in this study, is another important factor that would undoubtedly influence the results and warrants further exploration.

18.6 CONCLUSION

We investigated the potential moderating effect of mediated touch on social judgments. We explored how a mediated hug from a remote individual influences the impressions formed from hearsay information about a third person provided. Our findings indicate that a mediated hug reduces negative inferences in the recall of information about a target person. We also found a negative correlation between this reduction in negativity and preference for the target person. One possible mechanism behind these effects might be stress reduction facilitated by a mediated hug, which moderates the formation of negative impressions about a third party. We believe that using mediated hugs can temper the dissemination of negative information in telecommunication contexts.

ACKNOWLEDGMENT

This work was supported by JST the Exploratory Research for Advanced Technology (ERATO), the ISHIGURO Symbiotic Human-robot Interaction Project (Project Number: JPMJER1401), and partially by JST Core Research for Evolutional Science and Technology (CREST) Grant Number JPMJCR18A1, Japan.

REFERENCES

[1] Sakamoto D, Kanda T, Ono T, Ishiguro H, Hagita N. Android as a telecommunication medium with a human-like presence. In: *Human-Robot Interaction (HRI), 2007 2nd ACM/IEEE International Conference on IEEE.* 2007. p. 193–200.

[2] Tanaka K, Nakanishi H, Ishiguro H. Physical embodiment can produce robot operator's pseudo presence. In: *Investigating Human Nature and Communication through Robots.* 2017. p. 27.

[3] Haans A, IJsselsteijn W. Mediated social touch: A review of current research and future directions. *Virtual Reality.* 2006;9(2–3):149–159.

[4] Huisman G. Social touch technology: A survey of haptic technology for social touch. *IEEE Transactions on Haptics.* 2017;10(3):391–408.

[5] Gallace A, Spence C. The science of interpersonal touch: An overview. *Neuroscience & Biobehavioral Reviews.* 2010;34(2):246–259.

[6] Van Erp JB, Toet A. Social touch in human-computer interaction. *Frontiers in Digital Humanities.* 2015;2:2.

[7] McGlone F, Wessberg J, Olausson H. Discriminative and affective touch: Sensing and feeling. *Neuron.* 2014;82(4):737–755.

[8] Suvilehto JT, Glerean E, Dunbar RI, Hari R, Nummenmaa L. Topography of social touching depends on emotional bonds between humans. *Proceedings of the National Academy of Sciences.* 2015;112(45):13811–13816.

[9] Fisher JD, Rytting M, Heslin R. Hands touching hands: Affective and evaluative effects of an interpersonal touch. *Sociometry.* 1976;39(4):416–421.

[10] Patterson ML, Powell JL, Lenihan MG. Touch, compliance, and interpersonal affect. *Journal of Nonverbal Behavior.* 1986;10(1):41–50.

[11] Burgoon JK. Relational message interpretations of touch, conversational distance, and posture. *Journal of Nonverbal Behavior.* 1991;15(4):233–259.

[12] Burgoon JK,Walther JB, Baesler EJ. Interpretations, evaluations, and consequences of interpersonal touch. *Human Communication Research.* 1992;19(2):237–263.

[13] Hornik J. Tactile stimulation and consumer response. *Journal of Consumer Research.* 1992;19(3):449–458.

[14] Crusco AH, Wetzel CG. The midas touch the effects of interpersonal touch on restaurant tipping. *Personality and Social Psychology Bulletin.* 1984;10(4):512–517.

[15] Goldman M, Fordyce J. Prosocial behavior as affected by eye contact, touch, and voice expression. *The Journal of Social Psychology.* 1983;121(1):125–129.

[16] Paulsell S, Goldman M. The effect of touching different body areas on prosocial behavior. *The Journal of Social Psychology.* 1984;122(2):269–273.

[17] Bailenson JN, Yee N, Brave S, Merget D, Koslow D. Virtual interpersonal touch: Expressing and recognizing emotions through haptic devices. *Human-Computer Interaction.* 2007;22(3):325–353.

[18] Cabibihan JJ, Zheng L, Cher CKT. Affective tele-touch. In: *International Conference on Social Robotics.* Springer. 2012. p. 348–356.

[19] Haans A, de Bruijn R, IJsselsteijn WA. A virtual midas touch? Touch, compliance, and confederate bias in mediated communication. *Journal of Nonverbal Behavior.* 2014;38(3):301–311.

[20] Huisman G, Frederiks AD, van Erp JB, Heylen DK. Simulating affective touch: Using a vibrotactile array to generate pleasant stroking sensations. In: *International Conference on Human Haptic Sensing and Touch Enabled Computer Applications.* Springer. 2016. p. 240–250.

[21] Erk SM, Toet A, Van Erp JB. Effects of mediated social touch on affective experiences and trust. *PeerJ.* 2015;3:e1297.

[22] Nakanishi H, Tanaka K, Wada Y. Remote handshaking: Touch enhances video-mediated social telepresence. In: *Proceedings of the 32nd Annual ACM Conference on Human Factors in Computing Systems.* ACM. 2014. p. 2143–2152.

[23] Ogawa K, Nishio S, Koda K, Balistreri G, Watanabe T, Ishiguro H. Exploring the natural reaction of young and aged person with telenoid in a real world. *Journal of Advanced Computational Intelligence and Intelligent Informatics.* 2011;15(5):592–597.

[24] Morris D. *Manwatching: A field guide to human behavior.* Abrams. 1977.

[25] Hall ET, Birdwhistell RL, Bock B, Bohannan P, Diebold Jr AR, Durbin M, Edmonson MS, Fischer J, Hymes D, Kimball ST, et al. Proxemics [and comments and replies]. *Current Anthropology.* 1968;9(2/3):83–108.

[26] Forsell LM, Åström JA. Meanings of hugging: From greeting behavior to touching implications. *Comprehensive Psychology.* 2012;1:2–17.

[27] Grandin T. My experiences as an autistic child and review of selected literature. *Journal of Orthomolecular Psychiatry.* 1984;13(3):144–174.

[28] Edelson SM, Edelson MG, Kerr DC, Grandin T. Behavioral and physiological effects of deep pressure on children with autism: A pilot study evaluating the efficacy of grandin's hug machine. *American Journal of Occupational Therapy.* 1999;53(2):145–152.

[29] Mueller F, Vetere F, Gibbs MR, Kjeldskov J, Pedell S, Howard S. Hug over a distance. In: *Chi'05 Extended Abstracts on Human Factors in Computing Systems.* ACM. 2005. p. 1673–1676.

[30] Teh JK, Tsai Z, Koh JT, Cheok AD. Mobile implementation and user evaluation of the huggy pajama system. In: *Haptics Symposium (Haptics), 2012 IEEE.* IEEE. 2012. p. 471–478.

[31] Morikawa O, Hashimoto S, Munakata T, Okunaka J. Embrace system for remote counseling. In: *Proceedings of the 8th International Conference on Multimodal Interfaces.* ACM. 2006. p. 318–325.

[32] DiSalvo C, Gemperle F, Forlizzi J, Montgomery E. The hug: An exploration of robotic form for intimate communication. In: *Robot and Human Interactive Communication, 2003. Proceedings. Roman 2003. the 12th IEEE International Workshop on.* IEEE. 2003. p. 403–408.

[33] Sumioka H, Nakae A, Kanai R, Ishiguro H. Huggable communication medium decreases cortisol levels. *Scientifc Reports.* 2013;3:3034.

[34] Kuwamura K, Sakai K, Minato T, Nishio S, Ishiguro H. Hugvie: A medium that fosters love. In: *Ro-man, 2013 IEEE. IEEE.* 2013. p. 70–75.

[35] Nakanishi J, Kuwamura K, Minato T, Nishio S, Ishiguro H. Evoking affection for a communication partner by a robotic communication medium. In: *the 1st International Conference on Human-Agent Interaction (IHAI 2013).* 2013.

[36] Takahashi H, Ban M, Osawa H, Nakanishi J, Sumioka H, Ishiguro H. Huggable communication medium maintains level of trust during conversation game. *Frontiers in Psychology.* 2017;8:1862.

[37] Higgins ET, Rholes WS. "Saying is believing": Effects of message modi_cation on memory and liking for the person described. *Journal of Experimental Social Psychology.* 1978;14(4):363–378.

[38] Higgins ET. "Saying is believing" effects: When sharing reality about something biases knowledge and evaluations. In LL Thompson, JM Levine, & DM Messick (Eds.), *Shared cognition in organizations: The management of knowledge.* 1999;1:33–48. Lawrence Erlbaum Associates Publishers.

[39] Echterhoff G, Higgins ET, Groll S. Audience-tuning effects on memory: The role of shared reality. *Journal of Personality and Social Psychology.* 2005;89(3):257.

[40] Ackerman JM, Nocera CC, Bargh JA. Incidental haptic sensations inuence social judgments and decisions. *Science.* 2010;328(5986):1712–1715.

[41] Amir N, Beard C, Bower E. Interpretation bias and social anxiety. *Cognitive Therapy and Research.* 2005;29(4):433–443.

[42] Whitcher SJ, Fisher JD. Multidimensional reaction to therapeutic touch in a hospital setting. *Journal of Personality and Social Psychology.* 1979;37(1):87.

[43] Shermer M. A bounty of science. *Scientific American.* 2004;290(2):33–33.

[44] Ditzen B, Neumann ID, Bodenmann G, von Dawans B, Turner RA, Ehlert U, Heinrichs M. Effects of different kinds of couple interaction on cortisol and heart rate responses to stress in women. *Psychoneuroendocrinology.* 2007;32(5):565–574.

[45] Feldman R, Singer M, Zagoory O. Touch attenuates infants' physiological reactivity to stress. *Developmental Science.* 2010;13(2):271–278.

[46] Morhenn V, Beavin LE, Zak PJ. Massage increases oxytocin and reduces adrenocorticotropin hormone in humans. *Alternative Therapies in Health and Medicine.* 2012;18(6):11.

CHAPTER 19

Multi-modal Interaction through Anthropomorphically Designed Communication Medium to Enhance the Self-disclosures of Personal Information

Nobuhiro Jinnai, Hidenobu Sumioka, Takashi Minato, and Hiroshi Ishiguro

Osaka University, Osaka, Japan

Advanced Telecommunications Research Institute International, Kyoto, Japan

RIKEN, Saitama, Japan

DOI: 10.1201/9781003384274-24

19.1 INTRODUCTION

Communication devices have revolutionized how we connect with others, simplifying making new friends or partners and maintaining relationships with friends and family. Unfortunately, numerous researchers have emphasized that interaction by communication devices weakens human relations compared to face-to-face interactions [1–3]. To address this issue, we proposed a human-like robotic avatar that represents a distant individual in a communication process. The tangible presence of such avatars enables users to feel the existence of remote individuals [4]. Prior studies involving a lifelike android, Geminoid, suggest that using such robots as telecommunication devices strengthens the felt presence of remote individuals compared to standard audio or video conversations [5]. Interestingly, even human-like robotic avatars with minimal human resemblance show great potential as communication mediums, promoting positive relationships between humans [6]. However, no research has yet investigated how prolonged use of human-like communication media influences the development of intimate relationships between users who are meeting for the first time.

In this study, we explore how individuals form relationships with strangers when they interact by human-like communication media or traditional mobile phones over a period of roughly one month. Our findings indicate that a human-like communication medium facilitates the development of strong relationships with unfamiliar individuals. This holds true even when the functions of this medium mirror those of a standard mobile phone, with the only exception being its soft, human-like physical appearance.

19.2 ELFOID: A HANDHELD ROBOTIC MEDIUM

Elfoid™ is a unique communication device designed to embody the presence of a remote person [7]. Its key characteristic includes its miniaturized humanoid design, which features a humanoid head, arms, and legs. Its skin is constructed from soft sponge-like material, offering a tactile experience that resembles human skin. Elfoid's prototype was intended as a replacement for conventional mobile phones. However, in our study, we utilized an Elfoid-shaped, mobile-phone cover to offer functionalities similar to those of a typical mobile phone; the primary differences are its soft body and humanoid appearance (Figure 19.1). Users can communicate with remote partners by holding the device in a hands-free manner during conversations.

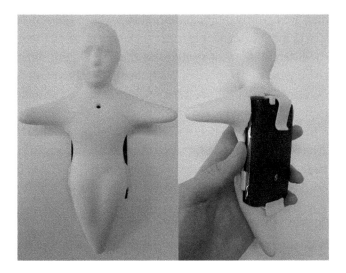

FIGURE 19.1   Elfoid: This version is equivalent to a mobile phone cover, which we placed over a standard mobile phone (Sony Ericsson Xperia mini) [18].

19.3 CURRENT STUDY

Prior studies demonstrate that Elfoid facilitates a stronger sense of human presence compared to devices that possess a mechanical appearance [8, 9]. Past research also suggests that people tend to become more intimate during face-to-face interactions compared to interactions mediated through audio or video [1–3]. Consequently, we hypothesize that Elfoid promotes a strong sense of togetherness, thereby allowing people to experience a closer connection. More specifically, we explore the long-term impact of human-like communication media on human relationships and hypothesize that interactions through Elfoid will promote better relationships than interactions through a traditional mobile phone. Our research addresses the following questions:

- Does interaction mediated by Elfoid enhance the amount of self-disclosure?

- How does the amount of self-disclosure change over the course of long-term Elfoid use?

We approach the first question by measuring the amount of self-disclosure during telecommunications using Elfoid or a conventional mobile phone.

For the second, we investigate the temporal changes in the self-disclosure levels of the participants. According to the social penetration theory [10, 11] and the early differentiation of relatedness theory [12, 13], different perspectives exist regarding the pace of relationship development. Although the former suggests that intimate relationships emerge through a gradual increase in self-disclosure, the latter posits that they develop rapidly through the initial attraction between strangers, as opposed to a gradual progress. Therefore, we anticipate observing either a gradual or rapid increase in the level of self-disclosure in conversations facilitated by Elfoid.

19.4 EXPERIMENT

19.4.1 Participants

Twelve individuals participated in this experiment, including four males and eight females, whose average age was 22.5 years. They were split into two groups, each comprised of three pairs (two male–female pairs and one female–female pair). The first group (the Elfoid group) conducted conversations through the Elfoid device from separate rooms at Osaka University. The second group (the Phone group) conversed by mobile phones (Figure 19.2). Before initiating the experiment, we ensured that the paired partners did not know each other. All participants provided written informed consent. This study received approval from the Ethics Committee of the Advanced Telecommunications Research Institute International (Kyoto, Japan).

19.4.2 Procedure

Each pair engaged in a ten-minute conversation twice a week, culminating in a total of ten conversations (i.e., 100 minutes) over a span of one month. The experiment was conducted in two distinct rooms. To prevent the pairs

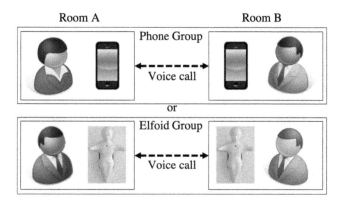

FIGURE 19.2 Experimental setting [18].

from encountering each other during the experiment, we requested that they arrive at different times.

In the experiment, participants were given either an Elfoid device or a mobile phone that was set to the hands-free mode and instructed to initiate a conversation. All the conversations were recorded. Apart from discussing topics related to religion or politics, conversation topics were unrestricted. Moreover, we asked the participants to address each other by their family names and prohibited them from asking for their partner's first name. This step was intended to discourage them from searching for their partners on social media and perhaps establishing contact outside of the experiment. After ten minutes, an experimenter entered one of the rooms and instructed one of the partners to end the conversation.

Participants completed a social skills inventory (SSI) scale [14] either before or after the experiment to measure their social skills.

19.4.3 Assessment of Social Skills

We evaluated the social skills of each participant using SSI, because social skills affect the establishment of close relationships with others. SSI has seven factors: emotional expressivity, emotional sensitivity, emotional control, social expressivity, social sensitivity, social control, and social manipulation. Among these factors, social expressivity refers to general verbal speaking skills and the ability to engage with others in social interactions [14] (Table 19.1).

19.4.4 Analysis of Self-disclosure in Conversation

We measured the evolution of intimate relationships by analyzing the amount of self-disclosure during conversations. Some researchers argue that its depth or intimacy increases as relationships become closer [10, 15]. Niwa and Maruno examined the conversation topics people engage in when they want to share more about themselves and developed a questionnaire

TABLE 19.1 Self-reported Copresence Scale (Q7 Was Added for This Study)

	Item
Q1	I did not want a deeper relationship with my interaction partner.
Q2	I wanted to maintain a sense of distance between us.
Q3	I was unwilling to share personal information with my interaction partner.
Q4	I wanted to make the conversation more intimate.
Q5	I tried to create a sense of closeness between us.
Q6	I was interested in talking to my interaction partner.
Q7	Do you want to know your conversation partner more?

for assessing the depth of self-disclosure [16]. This questionnaire includes 24 conversation topics, rated on a scale from 1 (I do not want to discuss this topic with the target person at all) to 7 (I am greatly interested in discussing this topic in detail with the target person). The questionnaire is sensitive to different degrees of self-disclosure, encompassing four levels: hobbies (level 1), difficult experiences (level 2), quirks (level 3), and shortcomings in personality traits and abilities (level 4). Each level comprises several topics (Table 19.2).

We utilized these topics to evaluate the self-disclosure of each participant in the conversations. We reviewed what each participant discussed in every session using the recorded video, and we determined whether

TABLE 19.2 Depth of Self-disclosure and Conversation Topics

Level	Item
Level 1	Your favorites (e.g., music, movie, and style of clothing)
	Ways to spend the weekend
	Something fun that happened recently
	Something about which you are enthusiastic recently
	Hobbies
	Events to which you are looking forward
	Something that you want to do
Level 2	Your experiences where someone helped you get out of a difficult situation
	Efforts you made to get out of a difficult situation
	Your method of overcoming bitter experiences
	Lessons that you learned the hard way
Level 3	Small faults that you have observed in yourself (unpunctuality, etc.)
	Some bad habits that you want to get over but cannot
	Minor faults that you get depressed thinking about
	Your experiences where you thought of yourself as something like a worthless being
	Your minor faults that others worry about
	Your minor faults that you worry about
Level 4	Some aspects of your characters that you hate
	Your experiences where you showed someone some aspects of your character that you hate
	Your abilities that you worry about
	Your experiences where you could not achieve a goal due to limitation of your ability
	Some abilities about which you have a sense of inferiority
	Your experiences where you felt disappointed due to the limitation of your abilities
	Your experiences where you hurt someone

the discussed topics were included in the list in Table 19.2. If a topic was included, it was categorized as a topic for self-disclosure made by the participant at the level to which the topic belongs. For example, if a participant discussed his/her hobbies with his/her partner, we classified it as a level 1 topic for self-disclosure. We often encountered several "sub-topics" related to self-disclosure within one topic. We counted each such example as a separate topic for self-disclosure if they were included on the list. For instance, if a participant talked about his/her favorite kind of music and then mentioned his/her favorite band, we considered it two separate topics for self-disclosure at level 1. Topics such as daily activities, lessons taken, the weather, or this study were not considered for self-disclosure because they were not on the list. We conducted such analysis for each participant across all sessions.

Using this analysis, we computed the number of topics at each level for each session to examine how the amount of self-disclosure varied over the month's term. We also calculated standardized values for the self-disclosure of each participant for each session to mitigate the effect of individual variation. We computed the total amount of self-disclosure from each participant across all the levels and all the sessions, as well as the total amount of each level of self-disclosure.

19.5 RESULT

19.5.1 Social Expressivity in SSI

The average social expressivity scores, as measured by the SSI, were 2.88 (SD: 0.54) for the Elfoid group and 3.10 (SD: 0.34) for the Phone group. According to the Shapiro-Wilk test, since the data from the Phone group did not exhibit a normal distribution, we applied the Mann-Whitney U-test. The results showed no significant difference between the two groups ($W(5)=12$, $p=.38$, ES: $r=.49$). This suggests that the participants in both groups possessed similar levels of social skills.

19.5.2 Amount of Self-disclosure

Figure 19.3a displays the average total amounts of self-disclosure for both groups. The participants in the Elfoid group exhibited an average of 17.17 instances (SD: 6.96) of self-disclosure, while those in the Phone group had an average of 7.67 (SD: 2.21). After confirming normality between the Elfoid and Phone groups using the Shapiro-Wilk test, we applied a Welch's t-test. The results indicated that the participants in the Elfoid group displayed significantly more self-disclosures than those in the Phone group ($t(5)=2.91$, $p=.027$, ES: $d=1.68$).

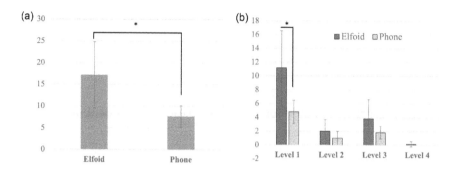

FIGURE 19.3 Average and standard deviation of the total amount of self-disclosure [18]. (a) Total amount of self-disclosure [18]. (b) Total amount of self-disclosure in each level [18].

We also compared the total amounts of self-disclosure at each level between the two groups (Figure 19.3b). On topics related to hobbies (level 1), participants in the Elfoid group exhibited 11.12 self-disclosures (*SD*: 5.40), while those in the Phone group made 4.83 self-disclosures (*SD*: 1.67). We found a significant difference between these groups ($t(5.954)=2.51$, $p=.046$, ES: $d=1.57$) using Welch's *t*-test after confirming the normality with a Shapiro-Wilk test. Although participants in the Elfoid group had more self-disclosures than those in the Phone group, we found no significant differences between the Elfoid group and the Phone group at the other levels, respectively: level 2: 2.00 (*SD*: 1.73) and 1.00 (*SD*: 1.00); level 3: 3.83 (*SD*: 2.79) and 1.83 (*SD*: 0.90); level 4: 0.17 (*SD*: 0.37) and 0.00 (*SD*: 0.0).

19.5.3 Temporal Changes of Amount of Self-disclosure

We also investigated the temporal changes of the amount of self-disclosure made by the participants. Two psychological theories (social penetration [10] and early differentiation of relatedness [12]) propose different perspectives regarding the speed of relationship development. Although the former concludes that intimate relationships develop through a gradual increase of self-disclosure, the latter argues that they instead progress quickly through the initial attraction between strangers. We next verified which theory is supported by the temporal change in the amount of self-disclosure.

Figure 19.4a,b shows the averages and standard deviations of the self-disclosure amounts made by each participant in each conversation session in the Elfoid and Phone groups. We found neither a gradual increase proposed by the social penetration theory nor a rapid increase proposed

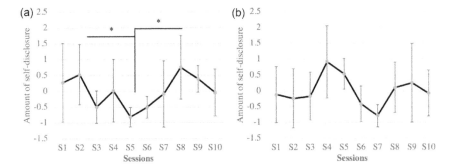

FIGURE 19.4   Changes in the amount of self-disclosure in Elfoid and Phone groups. Values are standardized ($M=0$, $SD=1$). Error bars represent standard deviations. *: $p<.05$ [18]. (a) Elfoid group and (b) Phone group.

by the early differentiation of relatedness theory. Instead, we observed a cyclic variation rather than a monotonic or rapid increase in both the Elfoid and Phone groups: the amount of self-disclosure first increased, then decreased, and finally increased again.

Therefore, we statistically verified the occurrence of such cyclic variations. We selected the session (S_{min}) that showed the minimum amount of self-disclosure and two sessions that showed the maximum amount before and after S_{min}. Thus, we selected the second, fifth, and eighth sessions in the Elfoid group and the fourth, seventh, and ninth sessions in the Phone group. We applied a repeated analysis of variance (ANOVA) to these three sessions in each group. In the Elfoid group, there was a significant main effect of sessions ($F(2)=4.54$, $p=.040$, $\eta_G^2=.42$). A post-hoc comparison with a modified sequentially rejective Bonferroni (MSRB) showed that participants in the Elfoid group expressed more self-disclosure in the second ($t(5)=3.54$, $p<.05$) and eighth ($t(5)=3.28$, $p<.05$) sessions than in the fifth session. On the other hand, in the Phone group, only the data between the seventh session and the fourth session approached significance ($F(2)=3.25$, $p=.08$, $\eta_G^2=.33$). The positions of the peaks and troughs corresponded to those in the amount of self-disclosure in level 1 (Figure 19.5).

19.6 DISCUSSION

Our results indicate that a communication medium with a human-like appearance increases the amount of self-disclosure, which measures intimacy [10, 15], even if it only has a human-like design and soft body, and all its other functions are identical as in standard mobile phones. Our results

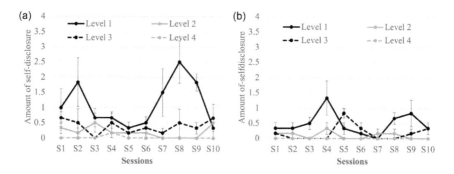

FIGURE 19.5 Changes of the amount of self-disclosure of each level in the Elfoid and Phone groups. Values are not standardized. Error bars represent standard errors [18]. (a) Elfoid group and (b) Phone group.

support the hypothesis that interaction mediated by human-like communication media facilitates more intimate relationships compared to typical communication media.

One possible explanation for this difference is that interacting through Elfoid mirrors more intimate, face-to-face interaction. Proxemics studies suggest that interpersonal distance represents human relationships [17]. Interestingly, while the Phone group participants placed their phones on a desk in the hands-free mode, five of six Elfoid group participants held it throughout the experiment. The sixth participant began holding it during the seventh session and continued to do so for the rest of the experiment. Two participants frequently touched Elfoid, indicating a misattribution of affinity toward it to their conversation partners due to its intimate distance (Figure 19.6). Further investigation is required to confirm this finding.

Another explanation might be that the participants in the Elfoid group felt a stronger sense of connectedness with their conversation partners. For instance, some Elfoid group participants treated their Elfoid as if it were human (Figure 19.7). Such behaviors, not observed in the Phone group, suggest that its anthropomorphic features invoke a sense of face-to-face interaction with a conversation partner.

Our findings also revealed that the development of an intimate relationship is influenced by changes in the amount of self-disclosure. For example, the amount of self-disclosure declined between certain sessions and then rose again, suggesting a cyclical self-disclosure pattern. This pattern was statistically confirmed in the Elfoid group and seemed to exist in the Phone group, suggesting that Elfoid accelerates this cyclical self-disclosure pattern.

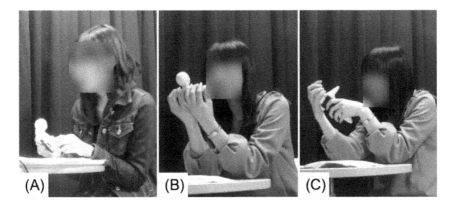

FIGURE 19.6   Touching behavior toward Elfoid. Participants (a) stroked its legs, (b) stroked its arms, and (c) held its head [18].

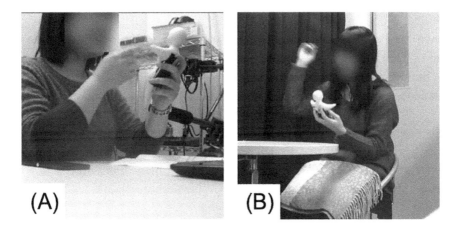

FIGURE 19.7   Gesture toward Elfoid. (a) Pointing gesture toward Elfoid. (b) Throwing motion [18].

No evidence was found to support the social penetration theory, possibly due to the short, one-month duration of our study. That theory often focuses on relationships that develop over the course of several months or even an entire year. Our results suggest that self-disclosure increases with the repetition of a cycle over longer periods. Our study demonstrated that a human-like communication medium increased self-disclosure between strangers after they first met, supporting the early differentiation of the relatedness theory. Perhaps the Elfoid group pairs developed a more intimate relationship than the Phone group pairs.

Our study does have several limitations. First, its sample size was relatively small. Second, while Elfoid does not have any actuators, its movement might induce a stronger feeling of presence. Finally, it's unclear which distinguishing factor of the device (appearance or softness) primarily contributed to our results. Future research will investigate these factors further and perhaps use a human-like communication device made of hard material.

19.7 CONCLUSION

We demonstrated that the long-term use of a human-like communication medium, known as Elfoid, elicited a greater degree of self-disclosure in conversations between unacquainted individuals compared to a standard mobile phone. We hypothesized that interactions through Elfoid would promote better relationships compared to interactions occurring by a traditional mobile phone. Our results confirmed that such a medium hastened the establishment of intimate relationships. Our research enriches the understanding of the impact of human-like communication media on human relationships and provides insights that might address various challenges in the field of telecommunications.

ACKNOWLEDGMENTS

This work was supported by JST CREST Grant Number JPMJCR18A1, Japan and JST, ERATO, Ishiguro Symbiotic Human-Robot Interaction Project (Grant Number: JPMJER1401). The authors thank Dr. Soheil Keshmiri for his valuable comments on this work.

REFERENCES

[1] C. O'Malley, S. Langton, A. Anderson, G. Doherty-Sneddon, and V. Bruce, "Comparison of face-to-face and video-mediated interaction," *Interacting with Computers*, Vol. 8, No. 2, pp. 177–192, 1996.

[2] G. M. Stephenson, K. Ayling, and D. R. Rutter, "The role of visual communication in social exchange," *British Journal of Social and Clinical Psychology*, Vol. 15, No. 2, pp. 113–120, 1976.

[3] K. Murase, "Distance counseling, counseling approaches, and state anxiety," *The Japanese Journal of Personality*, Vol. 14, No. 3, pp. 324–326, 2006 (in Japanese).

[4] E. Paulos and J. Canny, "Social tele-embodiment: Understanding presence," *Autonomous Robots*, Vol. 11, No. 1, pp. 87–95, 2001.

[5] D. Sakamoto, T. Kanda, T. Ono, H. Ishiguro, and N. Hagita, "Android as a telecommunication medium with a human-like presence," *Proc. of 2nd ACM/IEEE Int. Conf. on Human-Robot Interaction (HRI)*, pp. 193–200, 2007.

[6] H. Sumioka, S. Nishio, T. Minato, R. Yamazaki, and H. Ishiguro, "Minimal human design approach for sonzai-kan media: Investigation of a feeling of human presence," Cognitive Computation, Vol. 6, No. 4, pp. 760–774, 2014.

[7] T. Minato, H. Sumioka, S. Nishio, and H. Ishiguro, "Studying the influence of handheld robotic media on social communications," *Proc. RO-MAN 2012 Workshop on Social Robotic Telepresence*, pp. 15–16, 2012.

[8] H. Sumioka, K. Koda, S. Nishio, T. Minato, and H. Ishiguro, "Revisiting ancient design of human form for communication avatar: Design considerations from chronological development of dogu," *Proc. of IEEE Int. Symp. on Robot and Human Interactive Communication*, pp. 726–731, 2013.

[9] K. Tanaka, H. Nakanishi, and H. Ishiguro, "Physical embodiment can produce robot operator's pseudo presence," *Frontiers in ICT*, Vol. 2, p. 8, doi: 10.3389/fict.2015.00008, 2015.

[10] I. Altman and D. A. Taylor, *Social Penetration: The Development of Interpersonal Relationships*, Holt, Rinehart & Winston, 1973.

[11] D. A. Taylor, "Some aspects of the development of interpersonal relationships: social penetration processes," *Journal of Social Psychology*, Vol. 75, pp. 79–90, 1968.

[12] J. H. Berg and M. S. Clark, "Differences in social exchange between intimate and other relationships: Gradually evolving or quickly apparent?," V. J. Derlega and B. A. Winstead (Eds.), *Friendship and Social Interaction*, pp. 101–128, Springer, 1986.

[13] J. H. Berg, "Development of friendship between roommates," *Journal of Personality and Social Psychology*, Vol. 46, No. 2, pp. 346–356, 1984.

[14] R. E. Riggio, "Assessment of basic social skills," *Journal of Personality and Social Psychology*, Vol. 51, No. 3, pp. 649–660, 1986.

[15] T. H. Reis, P. Shaver et al., "Intimacy as an interpersonal process," S. Duck, D. F. Hay, S. E. Hobfoll, W. Ickes, and B. M. Montgomery (Eds.), *Handbook of Personal Relationships*, Vol. 24, No. 3, pp. 367–389, 1988.

[16] S. Niwa and S. Maruno, "Development of a scale to assess the depth of self-disclosure," *The Japanese Journal of Personality*, Vol. 18, No. 3, pp. 196–209, 2010 (in Japanese).

[17] E. T. Hall, *The Hidden Dimension*, Doubleday & Co., 1966.

[18] N. Jinnai, H. Sumioka, T. Minato, and H. Ishiguro, "Multi-modal interaction through anthropomorphically designed communication medium to enhance the self-disclosures of personal information," *Journal of Robotics and Mechatronics*, Vol. 32, No. 1, pp.76–85, doi: 10.20965/jrm.2020.p0076, 2020.

SECTION 6

Applications of Social Touch Interaction

A Minimal Design of a Human Infant Presence

A Case Study toward Interactive Doll Therapy for Older Adults with Dementia

Hidenobu Sumioka, Nobuo Yamato, Masahiro Shiomi, and Hiroshi Ishiguro

Advanced Telecommunications Research Institute International, Kyoto, Japan

The Japan Advanced Institute of Science and Technology, Ishikawa, Japan

Osaka University, Osaka, Japan

20.1 INTRODUCTION

Dementia is one primary cause of dependency and disability among older adults and significantly impacts them and their families, caregivers, and society. Approximately 40%–50% of individuals with dementia suffer from cognitive, psychological, and behavioral problems, all of which are collectively

DOI: 10.1201/9781003384274-26

labeled as the Behavioral and Psychological Symptoms of Dementia (BPSD). They include hallucinations, depression, and agitation [1]. Since these symptoms demand more attention from caregivers, their burden is increased as well as the care costs. Thus, reducing BPSD represents a significant social challenge. Although pharmacological interventions are often used for this purpose, non-pharmacological interventions are preferred to avoid potential side effects [2]. Many attempts related to non-pharmacological interventions have been addressed.

Doll therapy, which typically involves providing a human baby doll to seniors with dementia [3], is a type of non-pharmacological intervention that utilizes simulated social stimuli. Reports indicate that seniors with dementia engage in various caregiving activities with such dolls, including holding, talking, feeding, cuddling, and dressing them. This interaction not only enhances their engagement with others but also reduces problematic behaviors [4].

Although traditional doll therapy employs baby dolls without any interactive features, robotic technology can enhance interactions with older adults. For instance, Babyloid [5], which is a practical application of interactive doll therapy, has several functions, such as expressing emotions and body movements and vocalizing infant voices. However, robots with simpler mechanisms may also positively impact seniors with dementia.

In this study, we apply a minimal human design approach to an interactive baby robot to create a positive interaction for seniors with dementia by just expressing the minimum elements of human-like features and stimulating the user's imagination to supplement the missing information. Based on a minimal human design approach, we developed HIRO and investigated whether it induces longer interaction with dementia seniors than a baby robot with a face.

20.2 MINIMAL DESIGN OF HUMAN INFANT

Figure 20.1 shows a prototype of our minimal design of a human infant, HIRO (W210×D165×H300 mm and 610 g). Its ABS control module, which is covered with a polyester fabric, includes a computer, a three-axis accelerometer, and a speaker. The module can be removed from its back. Although HIRO is also equipped with a microphone and a touch sensor, we did not use them in this study since our purpose was to explore minimal requirements.

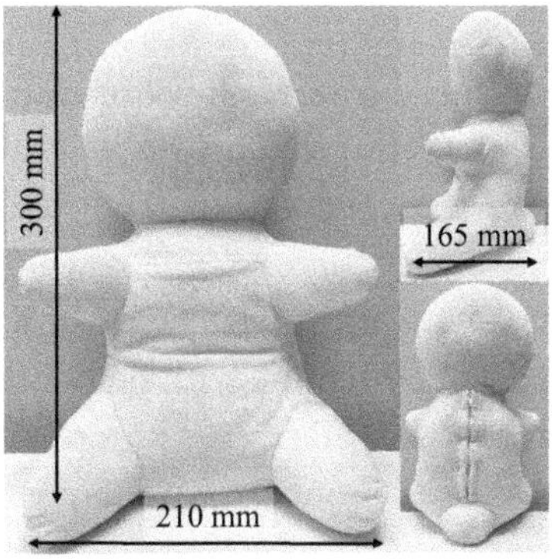

FIGURE 20.1 Minimal design of a human infant for interactive doll therapy, HIRO [15].

HIRO's design resembles a human baby: a distinct head, a torso, and limb sections, although no facial features. We did not design them because the mismatch between facial expressions and vocal tones can complicate the perception of emotion [6]. Furthermore, since we previously observed that older adults imagine facial expressions during interactions with Telenoid [7], we deliberately excluded any such facial features from HIRO.

To compensate for the lack of visual cues, we amplified HIRO's human-like qualities through auditory information. We recorded the voice of a one-year-old toddler and segmented it into 91 distinct voice patterns. A male college student, who was unaware of the study's purpose, classified these patterns into four emotional categories: positive (20 patterns), weakly positive (25 patterns), weakly negative (17 patterns), and negative (29 patterns). We also included three different types of babbling sounds in the weakly positive category, such as "pa-pa." The remaining sounds comprised various laughing and crying patterns.

The design of the interaction between the seniors and the robots is crucial. Here, we describe the details of HIRO's speech generation process. It has an internal emotional state that basically changes based on its sensor data derived from a three-axis accelerometer. Its emotional state is more often positive during greater interaction with a senior and less positive during less interaction. For example, HIRO often laughs when a user lifts or rocks it, but it often cries when a user places it on their legs or merely talks to it without moving it around.

20.3 EXPERIMENT

Since no study has introduced a faceless robot into nursing homes, we identified three critical research questions (RQs) that must be addressed before applying HIRO to practical doll therapy:

RQ1: Will a robot built on the minimal design of a human infant (like HIRO) be accepted by seniors with dementia?

RQ2: If HIRO is accepted, will it be received more positively than a baby robot that possesses more detailed body representations and facial expressions?

RQ3: Does HIRO induce different interaction patterns than a baby robot with facial features?

These research questions guided our investigation into HIRO's potential effectiveness and suitability as a therapeutic tool for seniors with dementia and set the stage for further exploration of the impact and the role of minimal design in human–robot interaction within elderly care contexts.

20.3.1 Participants

Our experiment was conducted in an elderly nursing home with 21 senior participants, 18 of whom were women. The average age of the participants was 86.6 years (*SD*: 5.4). The average level of the required long-term care (care level) for the participants was 3.38, ranging from levels 2 to 5, based on Japanese government guidelines. The care level determines the extent of attention required by a patient. For example, a senior at care level 2 may need partial assistance with daily activities such as personal hygiene and bathing, whereas a senior at care level 5 requires comprehensive care that encompasses every aspect of life. This level may also manifest numerous anxiety behaviors and a general decline in cognitive comprehension.

From among a group who regularly engaged in events held at the facility (including seniors who had difficulty speaking), the nursing home staff selected participants whose interactions with the robot would be observed. All the participants and their families were thoroughly informed about the experiment and provided signed informed consent forms. The experiment was conducted with the approval of the ATR Ethics Committee and carried out with permission from the participants' primary care physicians.

20.3.2 Procedure

The participants were randomly divided into two groups: one provided with HIRO (no-face group), consisting of 11 participants (including two males), and another given a robot with a more human-like appearance (face group), consisting of ten participants (including one male). Figure 20.2 illustrates the distinctions between the two robots. All the participants were parents who had raised more than one child, except for one participant in each group who did not raise any children. The detailed demographic information for each group is provided in Table 20.1.

Our study investigated whether the participants would continue to hold and positively interact with their baby robot for a five-minute duration after receiving it from a familiar staff member. This five-minute interval was suggested by the nursing staff as a representative minimum interval to alleviate their workload.

The experiment was conducted in the participants' private rooms, accommodating individual mobility needs such as sitting in chairs, remaining in wheelchairs, or staying in beds. The following is a detailed procedure of the experiment. First, a staff member and an experimenter entered a participant's room, introduced the robot, and handed it to the

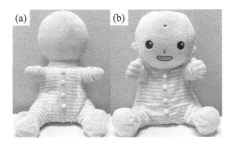

FIGURE 20.2   (a) HIRO in the no-face group and (b) a baby robot with a face in the face group [15].

TABLE 20.1   Demographic Information of Participants. F, M, and O Indicate Participant IDs

Group	Age (*SD*)	Care Level (*SD*)	Sp[a]	Be[b]
No-face	86.3 (4.21)	3.36 (0.81)	M, O	M
Face	86.9 (6.62)	3.4 (0.97)	F	F

[a] Participants with speech difficulty.
[b] Bedridden participants.

participant. The experimenter set up a video camera and left the room. About a minute later, the staff member received a call from the experimenter and explained to the participant that she needed to assist the experimenter, asking if the participant could care for the baby robot. To prevent undue pressure to comply, half of the participants in each group were accompanied by a nursing staff member, who stayed passively in the room. During this phase, the experimenter (and a staff member, if applicable) monitored the situation from outside. The interaction between the senior and the baby robot continued for the five-minute interval or until the participant lost interest (such as dropping the robot or falling asleep) or sought assistance from the staff. Then, the experimenter (and staff, if she had left) returned to the room, thanked the participant, retrieved the robot and the video camera, and exited. After the experiment, staff members were interviewed about their reactions to the interactions, especially noting any changes in their patients' behavior or states. We also collected the robots' emotional state histories during the interaction phase.

20.3.3 Evaluation

We investigated our above three research questions (RQ1, RQ2, and RQ3) through a combination of statistical analyses and observations. For RQ1, we analyzed whether the participants accepted the robot by holding it for the entire five minutes. The acceptance percentage was calculated for both groups. A one-sample proportion test assessed whether the percentage was higher than chance level for each group. For RQ2, two independent coders identified and encoded caring behaviors (e.g., caressing and hugging) and non-caring behaviors (e.g., just holding). A kappa coefficient of 0.79 indicated substantial agreement. We evaluated the duration of the caring behaviors using a Welch's t-test after confirming normality with the Shapiro-Wilk test. For RQ3, we analyzed the proportion of the emotional states selected by the robots and how they affected the interaction times, evaluated the difference in their emotional states between the two groups, and examined the correlations between the robots' emotional states and the participants' caring behaviors. Depending on the data's normality, a Welch's t-test or a Mann-Whitney's U test was used. We investigated with Pearson's correlation the linear relationships between the robots' emotional states and the participants' behaviors.

20.4 RESULT

The results of the investigation reveal insights into how the participants in an elderly nursing home interacted with robots of different appearances (with or without a face) over a five-minute period. Regardless of the group, most accepted their robot and continued to hold it for the entire five minutes (60% in the face group and 81.8 in the no-face group). The no-face group's acceptance was significantly higher than the chance level (50%) ($\chi^2=4.45$, $p=.017$), although the face group's acceptance was not ($\chi^2=.4$, $p=.26$). No significant difference was found between the groups ($\chi^2=1.22$, $p=.27$, 95% CI=[−0.16, 0.60]).

Most participants demonstrated some form of caring behavior, reflecting positive attitudes towards the robots. We analyzed the caring behaviors by separating them into verbal caring behaviors (talking and singing to the robot) and non-verbal ones. Figure 20.3 shows the total duration of the caring behaviors exhibited by each participant. The letters in this figure denote the participant IDs. The average duration of the verbal caring behaviors was 70.8 seconds (SD: 91.1) in the face group and 60.8 seconds (SD: 67.8) in the no-face group, showing no significant difference ($t=.23$, $p=.82$, $d=.13$). Similarly, there was no significant difference between the face group (191.0 seconds, SD: 74.6) and the no-face group (157.7 seconds, SD: 87.1) ($t=.79$, $p=.44$, $d=.41$).

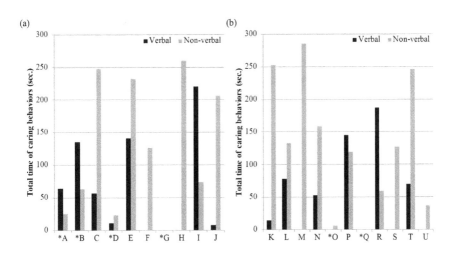

FIGURE 20.3   Total time of caring behaviors presented by each participant during interaction in (a) face group and (b) no-face group. Letters denote participant IDs. * shows participants who refused the robot [15].

We also examined the robots' emotional states to further understand the participants' engagement and interaction with them across the two groups. Figure 20.4 shows the average ratio of the emotional states of the robots over a five-minute period for each group. An asterisk indicates participants who rejected the robot for the entire five minutes. We found no distinctive differences between the groups or whether they interacted with it for the entire period. The emotional states were analyzed for positive, negative, weakly positive, and weakly negative conditions. For the negative states, the no-face group tended to have more negative states than the face group, although this was not significant (t (18.7)$=-1.80$, $p=.089$, $d=.79$). We found no significant difference between the groups for the weakly negative and weakly positive states (weakly negative: $W=63$, $p=.60$, $r=.17$; weakly positive: $W=68$, $p=.39$, $r=.28$). There was also no significant difference between the groups for the positive states ($t(16.1)=1.23$, $p=.24$, $d=.55$).

Finally, we conducted a correlation analysis to investigate any underlying relationships between the robots' emotional states and the participants' verbal/non-verbal caring behaviors. Pearson's correlation examined the relationship between the robots' emotional states and the total time of the participants' verbal/non-verbal caring behaviors. The analysis was performed for all the participants who accepted the robot in both groups, given the small sample size within each group. The results are shown in Table 20.2. A negative moderate correlation was identified ($r=-0.47$, $p=.03$) between the robots' negative states and the participants' verbal caring behaviors, showing that participants talked to their robot less often when it expressed more negative emotions like crying. We found a positive moderate correlation between the robots' weak positive state and the

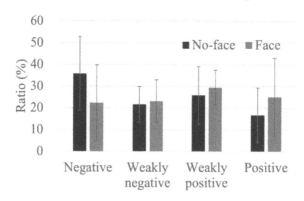

FIGURE 20.4 Average ratio of robot's emotional states during interaction [15].

TABLE 20.2   Correlation Coefficients between Caring Behavior and Robot's
State *: p < .05, †: p < .10

Caring	Robot's Internal State			
Behavior	**Negative**	**Weakly Neg.**	**Weakly Pos.**	**Positive**
Verbal	−0.47*	−0.28	0.44*	0.40†
Non-verbal	0.15	0.01	−0.32	0.05

participants' verbal caring behavior ($r=0.44$, $p=.046$), indicating that they talked to their robot more frequently when it expressed weak positive emotions. There were no significant correlations for other relationships. However, a trend approaching significance was noted for a positive correlation between verbal behavior and a robot's positive state ($r=0.40$, $p=.070$).

20.5 DISCUSSION

In this section, we adopted a minimalist design approach for creating an interactive doll named HIRO to explore the essential characteristics required for a robot to be effective in interactive doll therapy. We found that elderly participants accepted it and displayed positive attitudes, supporting RQ1. With RQ2 and RQ3, we did not find any difference between the two groups in terms of the absence of facial features and the robot's responses, rejecting both RQ2 and RQ3. Since our experiment lasted only five minutes, its impact may not have been significant. We need to test longer and multiple interactions.

Another possibility is a cultural dimension. It has been posited that Japanese people prioritize auditory cues over visual ones, especially when both are presented concurrently [8]. This might suggest that auditory interactions facilitated by robots are paramount for older Japanese adults with dementia. On the other hand, HIRO's simplistic design might resonate less with Europeans, who are perceived to lean more towards visual cues.

The influence of a human infant's voice on seniors was a salient feature in our study, with notable effects on their behavior. For this investigation, we utilized a recording of an actual human infant. However, the significance of human-like voice quality for enhancing the interaction between a robot and seniors remains an open question. Future research must explore whether similar outcomes are attainable with low-fidelity or synthesized speech.

In the realms of animal therapy and robot therapy [9,10], both have been accepted as mediators for communication among seniors and others.

This phenomenon was also observed in our experiment. During the study, when a nursing staff member was present in a participant's room, the participant frequently talked with her about the robot. However, since our experimental design explicitly requested staff members to refrain from active interaction with the participants, HIRO's potential as a communication facilitator between staff members and participants was not fully realized. As future work, we will verify our robot's effectiveness in a group situation where caregivers or seniors are present.

Although this study offers valuable insights, it is not without its limitations. Foremost is its relatively small sample size, coupled with an imbalance in the degree of dementia across the participant groups. Future investigations must separately gauge the participants' dementia levels using such objective metrics as the Severe Mini Mental State Examination (sMMSE). An objective assessment is also warranted that determines whether HIRO genuinely mitigates problem behaviors. Given pronounced individual variations, subsequent research should employ a within-subject design for more robust validation.

Another potential confounder is the pronounced gender bias within the participant pool, where male participants were notably underrepresented. This reflects demographic trend, where nearly two-thirds of dementia sufferers are women [11]. The gender imbalance is not unique to our study, as evidenced by previous research [12–14]. To ascertain the impact on male seniors with dementia, we must extend our study across multiple nursing homes and enhance the representation of male participants.

Finally, our assessment of HIRO's acceptance was predicated on a five-minute interaction with seniors afflicted with dementia. Although this duration aligns with other psychological studies and was chosen to minimize the workload of the care staff [12,14], it is relatively brief in the context of human–robot interaction research. Future inquiries must delve into the optimal length of engagement with a robot and the frequency of such interactions to ascertain HIRO's efficacy and its potential to alleviate caregivers' workloads.

20.6 CONCLUSION

We introduced a minimal design to develop a baby-sized robot called HIRO with abstract features for interactive doll therapy. Emotional interaction was achieved with the sounds of an actual baby. Our field study showed that seniors with dementia accepted HIRO, although we must expand the sample size and rigorously assess whether it reduces behavioral problems. We believe that HIRO will help improve their mental health and their quality of life in nursing homes.

ACKNOWLEDGMENT

This work was supported by JST CREST Grant Number JPMJCR18A1, Japan.

REFERENCES

[1] Cerejeira, J., Lagarto, L., and Mukaetova-Ladinska, E. B. (2012). Behavioral and Psychological Symptoms of Dementia. *Front. Neur.* 3, 73. doi:10.3389/fneur.2012.00073

[2] Azermai, M., Petrovic, M., Elseviers, M. M., Bourgeois, J., Van Bortel, L. M., and Vander Stichele, R. H. (2012). Systematic Appraisal of Dementia Guidelines for the Management of Behavioural and Psychological Symptoms. *Ageing Res. Rev.* 11, 78–86. doi:10.1016/j.arr.2011.07.002

[3] Mitchell, G. (2016). *Doll Therapy in Dementia Care: Evidence and Practice.* London: Jessica Kingsley Publishers.

[4] Mitchell, G., McCormack, B., and McCance, T. (2016). Therapeutic Use of Dolls for People Living with Dementia: A Critical Review of the Literature. *Dementia.* 15, 976–1001. doi:10.1177/1471301214548522

[5] Kanoh, M. (2015). A Robot as "Receiver of Care" in Symbiosis with People. *J. Soft.* 27, 193–201. doi:10.3156/jsoft.27.6_193

[6] de Gelder, B., and Vroomen, J. (2000). The Perception of Emotions by Ear and by Eye. *Cogn. Emot.* 14, 289–311. doi:10.1080/026999300378824

[7] Sumioka, H., Nishio, S., Minato, T., Yamazaki, R., and Ishiguro, H. (2014). Minimal Human Design Approach for Sonzai-Kan Media: Investigation of a Feeling of Human Presence. *Cogn. Comput.* 6, 760–774. doi:10.1007/s12559-014-9270-3

[8] Tanaka, A., Koizumi, A., Imai, H., Hiramatsu, S., Hiramoto, E., and de Gelder, B. (2010). I Feel Your Voice. *Psychol. Sci.* 21, 1259–1262. doi:10.1177/0956797610380698

[9] Shibata, T., Coughlin, J. F., and Coughlin, J. F. (2014). Trends of Robot Therapy with Neurological Therapeutic Seal Robot, PARO. *J. Robot. Mechatron.* 26, 418–425. doi:10.20965/jrm.2014.p0418

[10] Shibata, T. (2012). Therapeutic Seal Robot as Biofeedback Medical Device: Qualitative and Quantitative Evaluations of Robot Therapy in Dementia Care. *Proc. IEEE.* 100, 2527–2538. doi:10.1109/jproc.2012.2200559

[11] Fukawa, T. (2018). Prevalence of Dementia Among the Elderly Population in Japan. *Health Prim. Car.* 2 (4), 1–6. doi:10.15761/HPC.1000147

[12] Cohen-Mansfield, J., Marx, M. S., Dakheel-Ali, M., Regier, N. G., and Thein, K. (2010). Can Persons with Dementia Be Engaged with Stimuli? *Am. J. Geriatr. Psychiatry.* 18, 351–362. doi:10.1097/jgp.0b013e3181c531fd

[13] Kuwamura, K., Nishio, S., and Sato, S. (2016). Can We Talk Through a Robot as If Face-To-Face? Long-Term Fieldwork Using Teleoperated Robot for Seniors with Alzheimer's Disease. *Front. Psychol.* 7, 1066. doi:10.3389/fpsyg.2016.01066

[14] Pezzati, R., Molteni, V., Bani, M., Settanta, C., Di Maggio, M. G., Villa, I., et al. (2014). Can Doll Therapy Preserve or Promote Attachment in People with Cognitive, Behavioral, and Emotional Problems? A Pilot Study in Institutionalized Patients with Dementia. *Front. Psychol.* 5, 342. doi:10.3389/fpsyg.2014.00342

[15] Sumioka H, Yamato N, Shiomi M and Ishiguro H (2021) A Minimal Design of a Human Infant Presence: A Case Study Toward Interactive Doll Therapy for Older Adults With Dementia. *Front. Robot. AI.* 8, 633378. doi:10.3389/frobt.2021.633378

CHAPTER 21

Interactive Baby Robot for Seniors with Dementia

Long-term Implementation in Nursing Home

Nobuo Yamato, Hidenobu Sumioka, Hiroshi Ishiguro, Youji Kohda, and Masahiro Shiomi

The Japan Advanced Institute of Science and Technology, Ishikawa, Japan

Advanced Telecommunications Research Institute International, Kyoto, Japan

Osaka University, Osaka, Japan

21.1 INTRODUCTION

According to a White Paper on Aging (Cabinet Office, Government of Japan), in 2012, Japan had 4.62 million seniors with dementia, an amount that is expected to reach 6.75 million by 2025. Dementia affects not only the daily lives of the patients themselves but also their families and surrounding communities due to the Behavioral and Psychological Symptoms of Dementia (BPSD). This medical condition is characterized by problematic

DOI: 10.1201/9781003384274-27

behaviors such as verbal abuse, assault, and wandering as well psychological symptoms such as delusion and decreased motivation. Since caregivers must constantly watch over seniors with dementia, they experience
a heavy burden. Societies also face increased economic responsibilities.
Thus, BPSD reduction is a major social issue related to dementia [1, 2].

For addressing BPSDs in the early stages, non-pharmacological interventions that do not involve medication are recommended because of con-
cerns about the side effects of pharmacological approaches [3]. Recently,
robot therapy has started to use communication robots such as animal-like
robots and interactive robots, and its effectiveness is being scientifically
investigated [4].

As previously described in Section 6.1, we are focusing on a baby-like
interactive robot called HIRO inspired by doll therapy [5] to reduce BPSDs
and the burden on caregivers. Since the dolls used in doll therapy have
neither moving parts nor speech functions, we added a speech function
to a baby-like doll. Although we confirmed that seniors with dementia
can be actively involved with HIRO for five minutes, it remains unclear
whether such attitudes are sustained through long-term use. Nor have we
investigated the changes in the relationship between a robot and a senior
with dementia or caregivers when a robot is used for a long period of
time. Therefore, in this study, we introduce "Kamatte HIRO-chan" (here-
inafter called HIRO-chan), a new robot manufactured with the Vstone
Corporation based on HIRO (Figure 21.1) to a nursing home for two
weeks. We confirmed the effects and issues related to relationship building and improved the robot based on such issues. Next, we introduced our
improved system for about one month and concluded that it improved the
relationship between the robot and seniors with dementia and care workers. Based on our experiment results, we discuss the effectiveness of our
introduction of HIRO-chan from the viewpoints of seniors with dementia,
the nursing staff, and nursing care facilities.

21.2 BABY-LIKE INTERACTIVE ROBOT

In this study, we used HIRO-chan, which is based on HIRO described in
Section 6.1, a minimally designed robot whose form resembles a human
infant (Figure 21.1).

HIRO-chan is a baby-like soft doll about 30 cm tall. It contains only
a microcomputer connected to an acceleration sensor and a speaker
and has no driving mechanism. Its voice was recorded from an actual
one-year-old human infant, a feature that enables seniors with dementia to intuitively imagine emotional information such as crying and

FIGURE 21.1   Kamatte "Hiro-chan": a baby-like interactive robot [6].

laughing. We also collected more than 100 different voice patterns and classified them into five categories (negative, weakly negative, normal, weakly positive, and positive) based on the positive or negative emotions conveyed by them. One of the five categories is selected in accordance with HIRO-chan's emotional state, which is determined by its acceleration and the values of its sensors, and a corresponding audio file is played. For example, HIRO-chan frequently laughs when a user physically engages with it by lifting or rocking it; it cries otherwise. Its normal voice also includes babbling such as "pa-pa."

21.3 EXPERIMENT

We introduced HIRO-chan to a nursing home for about two weeks to investigate its long-term influence on seniors with dementia and caregivers. The experiment was conducted in a four-story nursing home. This study was approved by the Ethics Committee of the Advanced Telecommunications Research Institute International (ATR).

21.3.1 Preliminary Two-Week Experiment (Experiment 1)

21.3.1.1 Participants

This experiment involved nine residents (eight women and one man) living on the same floor of a nursing home whose ages ranged from 70 to 97 and averaged 84.9. The NPI-NH test results [7], which measure BPSD, ranged from 4 to 42 points with an average of 15 (the highest possible score is 120; higher scores indicate more severe BPSD). The Conversational Assessment of Cognitive Functioning through Daily Dialogue (CANDy) [8], which tests for dementia through conversation, ranged from 0 to 21, with an average of 9.67 (the highest possible score is 30, and a score of 6 or higher strongly indicates dementia). Eight staff members also participated in this experiment.

21.3.1.2 Experimental Procedure

To reduce the risks of COVID-19 infection, we prepared and assigned a HIRO-chan to each participant. Before the experiments, we taught through a remote meeting system all the facility's nursing staff how to use HIRO-chan and how to hand it to the participants to minimize differences of HIRO-chan's handling among the nursing staff. The nursing staff could assist any participants who wanted to design their own HIRO-chan by adding facial features to it.

The experiment proceeded as follows. First, a staff member entered a participant's room while cradling HIRO-chan like a baby: "Oh, look a baby. Why don't you hold it?" The nursing staff observed the participants' reactions and briefly recorded specific details on a form prepared in advance. The nursing staff were also asked to record (for each participant) any problems they encountered while operating HIRO-chan, their impressions of its recommended uses or possible improvements, and any suggestions for future functions.

21.3.1.3 Result

Hospitalization forced participants B and F to quit the experiment on its ninth and sixth days. Including these two, our participants used HIRO-chan for 6–12 days and directly interacted with it one to four times a day: an average of 13.6 times during the experiment. Active interactions with HIRO-chan such as hugging it, talking with it, and smiling at it were identified as manifestations of a positive, accepting attitude. Negative attitudes were characterized as when participants rejected HIRO-chan by saying "I don't want it" or throwing it without engaging with it. Neutral attitudes included situations where participants did not respond to its voice, left it by the bed, or concentrated on something else (e.g., eating). The percentage of each attitude per participant was then examined (Figure 21.2).

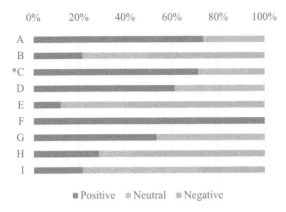

FIGURE 21.2 Percentage of each attitude toward HIRO-chan per participant: Letters on the vertical axis indicate participant IDs, and * indicates the sole male participant [6].

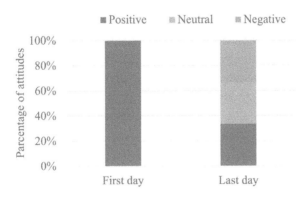

FIGURE 21.3 Percentage of each attitude of participants during first and last days of interaction with HIRO-chan [6].

Our results indicate that five participants (A, C, D, F, and G) had more positive attitudes throughout the experiment, one participant (I) had more neutral attitudes, and three participants (B, E, and H) had more negative attitudes toward HIRO-chan. In other words, participants with positive or neutral attitudes outnumbered those with negative attitudes. However, a comparison of the changes in the participants' attitudes based on their initial and final interactions showed that although all the participants had a positive attitude toward HIRO-chan on the first day, negative attitudes increased to about 30% by the end of the two-week period (Figure 21.3), indicating that their attitudes toward HIRO-chan had deteriorated.

21.3.1.4 Discussion

Although experiment 1 confirmed that HIRO-chan can be operated in a nursing home over a long period by nursing staff alone, its acceptability by seniors with dementia is not fully guaranteed for long-term use. Initially, HIRO-chan's novelty elicits a positive response from seniors with dementia. However, after repeated use, the number of participants who rejected it increased. The feedback obtained from the interviews with the nursing staff suggests that HIRO-chan's crying might have sparked this problem. Crying caused the seniors with dementia to stop interacting with HIRO-chan, a development that prevents the accelerometer from changing, which in turn causes HIRO-chan to enter a negative emotional state, leading to further crying. We assume that this pernicious cycle is responsible for the increased rejections of HIRO-chan by seniors with dementia.

Crying may also have negative effect on other residents (who were actually not involved in the experiment on the same floor) and nursing staff. In an interview, one staff member commented, "When HIRO-chan cries, I have to talk to the seniors who are sometimes caring for it," indicating that HIRO-chan's crying may have ironically increased the burden on nursing staff by requiring them to focus on seniors when they are actually not interacting with it.

From the above, we found that when HIRO-chan is introduced into a nursing home, the negative emotional expression of "crying" may negatively impact seniors with dementia who are also using the robot (at different times), as well as other residents and nursing staff. Some seniors were able to soothe HIRO-chan and make it laugh when it cried, suggesting a possible approach to control the robot's crying behavior based on feedback from seniors who used it. However, since the presence of HIRO-chan's crying function greatly complicated the easy and continuous eliciting of positive reactions from seniors with dementia, we modified its crying function and conducted another long-term experiment and investigated whether the positive attitude of seniors with dementia was maintained and how the modified HIRO-chan positively impacted the other residents and the nursing staff.

21.3.2 Long-term Experiment (Experiment 2)

As explained above, based on the results of Experiment 1, we developed HIRO-chan-S by modifying its voice. No changes were made to its exterior. The voices included in the positive, weakly positive, and normal categories were retained, although the negative and weakly negative voices were removed. We investigated whether the seniors with dementia continued to use HIRO-chan-S and whether our improved system addressed the issues identified in Experiment 1 based on the nursing staff's impressions.

21.3.2.1 Participants

Ten seniors (six women and four men) participated who lived on different floors of the same nursing home. Their ages ranged from 78 to 94 and averaged 87.9; their NPI-NH scores ranged from 0 to 13 and averaged 3.6; their CANDy scores ranged from 6 to 19 and averaged 11.8. Fourteen staff members also joined this experiment.

21.3.2.2 Experimental Procedure

Except for using HIRO-chan-S instead of HIRO-chan, we retained the same procedure as in Experiment 1. The staff again filled out questionnaires about their experiences with HIRO-chan-S. However, we omit the results here due to space limitations.

21.3.2.3 Results

Almost all the participants used HIRO-chan-S every day during the experiment. One participant (L) was unable to use it for three days due to illness, and two participants (M and S) missed one day due to conflicting caregiving schedules. Nine to ten staff members observed each participant. Figure 21.4 shows the percentage of each attitude by participant.

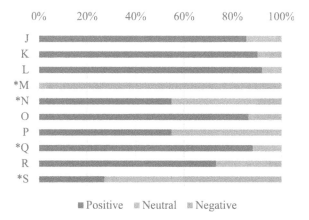

FIGURE 21.4   Percentage of each attitude per participant toward HIRO-chan-S [6].

Eight participants had mostly positive attitudes throughout the experiment, while two had mostly negative attitudes. No participant had a generally neutral attitude. Compared with the Experiment 1 results, the number of participants with positive attitudes increased, while those with negative and neutral attitudes decreased. Their attitudes throughout the experiment did not change toward HIRO-chan-S, compared with Experiment 1. Figure 21.5 shows the percentage of each attitude of all the participants during their first and last days with HIRO-chan-S. The attitude exhibited by the participants on the first day remained exactly the same. As in Experiment 1, there was no significant increase in their negative attitudes.

We divided the 13 comments from the nursing staff throughout Experiment 2 into three main categories: operational accidents (4 comments), voice volume (2), and additional functions (7). The operational accidents included problems that arose during the system's performance for participants K, L, and Q. For participants L and Q, HIRO-chan-S ran out of batteries once and we had to replace them. Participants K and L seemed fascinated by the clicking the volume button, which they pressed repeatedly, causing the robot to lose volume control. Therefore, the robot was replaced.

Both comments about volume pointed out that its voice was too low. The volume of course was fine for the authors and the nursing staff. However, some seniors with dementia struggled to hear. Three respondents suggested singing, and one each selected "sitting stably," "imitating the language of a senior," "responding to seniors," and "talking to seniors from three different age levels."

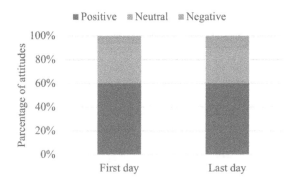

FIGURE 21.5 Percentage of each attitude of all participants during first day and last day with HIRO-chan-S [6].

21.4 DISCUSSION

21.4.1 Effects of Improved System on Seniors with Dementia

Our two-week introduction experiment (Experiment 1) confirmed that the number of subjects who rejected HIRO-chan increased, a result that identified a potential cause of such negative emotions expressed by HIRO-chan: crying. In a one-month introduction experiment using HIRO-chan-S (Experiment 2), where we removed its crying function, eight subjects continued to use HIRO-chan-S almost every day without rejecting it; one consistently rejected it, and the attitude of another changed from day to day. This suggests that the negative emotional expressions expressed by the robot in response to positive acceptance by the human side may have exacerbated confusion and anxiety on the human side. In this respect, HIRO-chan-S is basically in a good mood, and the program was modified to increase its cheerfulness when it is soothed, so the subjects calmly accepted it.

21.4.2 Passive Effect of Baby Interactive Robot on Other People

During the two-week introduction experiment, we confirmed that Hiro-chan's crying negatively impacted not only on the elderly people directly using the robot but also on the surrounding seniors and caregivers. A care staff member who participated in the experiment commented, "When HIRO-chan cries, I have to talk to the senior who use it and handle it together," which increases the need for care staff to pay more attention than usual to the senior. Unfortunately, this result completely undermines our objective: reducing the burden on caregivers. In our previous study, caregivers commented that they were soothed by listening to Hiro-chan's laughter [9], but if Hiro-chan's emotional expression is not properly controlled, it will both intensify the psychological burden on seniors who are directly involved and perhaps simultaneously increase the burden on the caregivers.

21.4.3 Reducing Caregiver Burdens

In this study, when a robot was introduced into a nursing home, not only the residents but also the caregivers were affected. Although a recent review concluded that such effects are positive, unfortunately, they can actually also be a factor that increases the mental stress and workload of caregivers [10]. Our previous study in Section 6.1 confirmed that seniors with dementia handle baby-like interactive robots, such as the one used in this study, like human babies without any instructions and call a caregiver when necessary. This means that seniors with dementia can satisfactorily handle the robot on their own without being taught by caregivers, suggesting that they do not need to hover over this robot as they do with other robots; they can leave and perform other tasks until they are summoned. However, in this study, to confirm any long-term changes in the relationship between seniors with dementia and the robot, the caregivers observed their relationships. In the future, we plan to ask the caregivers to perform another task and measure the time required for it to investigate the change in workload caused by our baby-like interactive robot.

21.4.4 Effects of Emotional Expression

Since dementia's core symptom is memory impairment, perhaps the participants will forget their involvement soon after a robot interaction. In fact, we confirmed that some participants had memory impairment from the observation records compiled by the caregivers in which the robot was referred to by different names on different days by the same senior. This seems inconsistent with the fact that the participants in Experiment 1 eventually rejected HIRO-chan, a result that this did not happen with HIRO-chan-S in Experiment 2. However, recent studies have shown that, although the neocortical functions that control knowledge and reason decline in seniors with dementia, the functions of the limbic system that control emotional memory and cognition are relatively preserved, and emotional memories such as pleasantness and unpleasantness are likely to remain [11]. This might explain why the impressions of the robot evoked by crying and laughing were remembered by the participants, even though they forgot their knowledge about it. We infer that HIRO-chan was avoided due to negative emotional memory, while HIRO-chan-S remained in use thanks to positive emotional memory.

However, negative emotional expressions from the robot did not always create a negative impact. In fact, for some seniors, the robot's crying may be associated with positive emotions. For example, as reported in our previous study [9], several subjects enjoyed comforting the crying robot. In an interview with caregivers from Experiment 2, one acknowledged that HIRO-chan-S was accepted by more people, although based on her experience with HIRO-chan, "some people were better off when they were compelled to tend to a crying baby." Therefore, to introduce a crying robot, the characteristics of seniors with dementia must be considered.

21.4.5 Limitations

In this study, since the number of seniors with dementia was small, we must increase the number of subjects to confirm the results. Although we found no gender difference in the effects in this study, the effects on men were not fully investigated due to the small number of male participants (14 women and five men for Experiments 1 and 2 combined). We must increase the overall number of participants as well as the proportion of male participants. We also analyzed the reports of caregiver's observations. However, we still need more detailed observations of the relationships between seniors with dementia and the robot. A future study might incorporate video recording and the analysis of senior reactions.

21.5 CONCLUSION

In this section, we conducted a two-week introduction experiment of a baby interactive robot at a nursing care facility, and based on its findings, we modified the robot's program and conducted another long-term one-month introduction experiment. Our baby interactive robot developed using a minimal design approach can be used for a long period of time if its emotional expressions are limited to those that give positive impressions and if it is operated in a nursing care facility. We also confirmed that, as a passive social medium, our baby interactive robot positively influences not only the users but also those around it. In the future, we will accelerate improvements by conducting longer-term introduction experiments and addressing the issues identified in them.

ACKNOWLEDGMENT

This research work was supported in part by JST CREST Grant Number JPMJCR18A1, Japan. We thank the staff and residents and their families of the Ryusei Fukushikai Special Care Nursing Home "Yume Paratiis" for supporting our experiment.

REFERENCES

[1] Livingston, G. et al.: Dementia prevention, intervention, and care, *The Lancet*, Vol. 390, pp. 2673–2734 (2017).

[2] Livingston, G. et al.: Dementia prevention, intervention, and care: 2020 report of the Lancet Commission, *The Lancet*, Vol. 396, pp. 413–446 (2020).

[3] Cerejeira, J., Lagarto, L. and Mukaetova-Ladinska, E. B.: Behavioral and psychological symptoms of Dementia, *Frontiers in Neurology*, Vol. 3, DOI: 10.3389/fneur.2012.00073. (2012).

[4] Broadbent, E., Stafford, R. and MacDonald, B.: Acceptance of healthcare robots for the older population: Review and future directions, *International Journal of Social Robotics*, Vol. 1, pp. 319–330 (2009).

[5] Mitchell, G., McCormack, B. and McCance, T.: Therapeutic use of dolls for people living with dementia: A critical review of the literature, *Dementia*, Vol. 15, No. 5, pp. 976–1001 (2016).

[6] Yamato, N., Sumioka, H., Ishiguro, H., Kohda, Y. and Shiomi, M.: Interactive baby robot for the elderly with Dementia-Realization of long-term implementation in nursing home, *IPSJ Transactions on Digital Practice*, Vol. 3, No. 4, pp. 14-27 (2022) (in Japanese).

[7] Wood, S. et al. The use of the neuropsychiatric inventory in nursing home residents: Characterization and measurement. *American Journal of Geriatric Psychiatry*, Vol. 8, 75–83 (2000).

[8] Oba, H., Sato, S., Kazui, H., Nitta, Y., Nashitani, T. and Kamiyama, A.: Conversational assessment of cognitive dysfunction among residents living in long-term care facilities. *International Psychogeriatrics*, Vol. 30, 87–94. DOI: 10.1017/S1041610217001740. (2018).

[9] Sumioka, H. et al.: A minimal design of a human infant presence: A case study toward interactive doll therapy for older adults with dementia, *Frontiers in Robotics and AI*, Vol. 8, DOI: 10.3389/frobt.2021.633378. (2021).

[10] Persson, M., Redmalm, D. and Iversen, C.: Caregivers' use of robots and their effect on work environment—A scoping review, *Journal of Technology in Human Services*, Vol. 40, pp. 1–27 (2021).

[11] Gineste, Y., and Pellissier, J.: Humanitude : Comprendre la vieillesse, prendre soin des Hommes vieux. Aramand Colin (2007).

CHAPTER 22

A Huggable Device Can Reduce the Stress of Calling a Stranger on the Phone for Individuals with ASD

Hidenobu Sumioka, Hirokazu Kumazaki, Taro Muramatsu, Yuichiro Yoshikawa, Hiroshi Ishiguro, Haruhiro Higashida, Teruko Yuhi, and Masaru Mimura

Advanced Telecommunications Research Institute International, Kyoto, Japan

Nagasaki University, Nagasaki, Japan

Keio University, Tokyo, Japan

Kanazawa University, Kanazawa, Japan

Osaka University, Osaka, Japan

DOI: 10.1201/9781003384274-28

273

22.1 INTRODUCTION

Recent innovations in communication devices, including mobile devices, have fundamentally altered human interaction [1]. In our daily life, mobile phones are essential for engaging with others, even strangers, in various contexts. However, for individuals with autism spectrum disorders (ASD), communication and language present significant challenges. Such individuals often struggle to talk to strangers on mobile phones, partly due to limited imaginative capabilities and heightened anxiety levels.

Stress and anxiety can hinder self-control and focus on communication partners [2], a problem exacerbated in ASD individuals due to their restricted self-regulation abilities. Research in psychology has found that tactile sensations enhance comfort during communication [3–6], and the physical presence of a conversation partner substantially influences a speaker's perception of her surroundings [7,8]. Interactions involving touch reduce stress, a well-known effect of interpersonal touch [3]. Consequently, interest is growing in simulating the psychological benefits of interpersonal touch through tactile sensations in communication devices.

Hugvie (Figure 22.1), a human-shaped pillow whose design replicates a hugging sensation, was developed to foster positive emotions like comfort and trust during phone conversations [9–12]. A connection and presence with a remote conversation partner can be enhanced by squeezing such a human-like form as Hugvie and hearing a voice close to the ears. However, ASD individuals often exhibit atypical responses to tactile stimuli [13–18], creating uncertainty about Hugvie's effectiveness for them during phone conversations with strangers. We believe that Hugvie may significantly

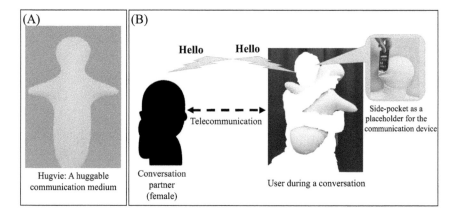

FIGURE 22.1 (a) Hugvie and (b) individual telecommunicating with a remote person while hugging a Hugvie [40].

alleviate stress in such interactions based on the communicating challenges faced by ASD individuals with unfamiliar people over the phone as well as the social dysfunction inherent in ASD.

Our primary aim is to investigate whether physical contact through hugging Hugvie alleviates stress in ASD individuals during phone conversations with strangers. Previous research concluded that hugging a Hugvie reduces stress in the general population [9], while ASD individuals often perceive textures as more pleasant than do controls [19]. Therefore, we hypothesize that Hugvie usage will similarly benefit ASD individuals who are preparing for phone conversations with unknown persons and utilized self-reporting techniques and salivary cortisol level measurements to assess how Hugvie affected their self-confidence and stress indicators.

22.2 MATERIALS AND METHODS

22.2.1 Participants

We recruited young adults diagnosed with ASD from Kanazawa University by displaying posters within institutions on campus. The Ethics Committee of Kanazawa University approved the study, which was conducted in accordance with both the institutional/national research committee's standards and the 1964 Declaration of Helsinki. All participants and their parents gave written informed consent after receiving a comprehensive explanation of the study's objectives and procedures. The selection criteria for participants were as follows: (1) an age range of 15–24; (2) an IQ of 70 or above; and (3) validation from seasoned psychiatrists concerning the ability to understand and accurately complete the informed consent forms, questionnaires, and experimental procedures. Even in the case of one individual whose IQ was below 70, experienced psychiatrists affirmed the ability to provide accurate written consent and comprehend the study's methodology.

Upon enrollment, all the participants were evaluated by trained psychiatrists and diagnosed with ASD, adhering to the criteria in the Diagnostic and Statistical Manual of Mental Disorders (DSM-5) [20] and the standardized guidelines in the Diagnostic Interview for Social and Communication Disorders (DISCO) [21], which assess early development and daily activities. This procedure enabled the interviewer to gauge the ASD-affected individuals' functionality in areas beyond social interaction and communication [22]. We confirmed with the Mini-International Neuropsychiatric Interview (M.I.N.I.) [23] that no participants had any other psychiatric disorders except ASD. We also administered the Autism Spectrum Quotient-Japanese version (AQ-J), which assesses ASD-specific behaviors and symptoms [24], the Wechsler Adult Intelligence Scale-Fourth Edition [25], which gauges

IQ, the Liebowitz Social Anxiety Scale (LSAS) [26], which quantifies the intensity of social anxiety symptoms, the ADHD Rating Scale (ADHD-RS) [27], which evaluates inattentive and hyperactive-impulsive symptoms, and the Adolescent/Adult Sensory Profile (AASP) [28], which assesses sensory processing in individuals aged 11 or older.

22.2.2 Procedure

The research was conducted as a crossover study in which participants engaged in a phone dialogue with an unfamiliar partner while either embracing a Hugvie equipped with a mobile phone (Hugvie session) or just utilizing a mobile phone (Phone session). The unknown phone partner was a 34-year-old female without any specialized training in communicating with individuals with ASD. She served as a consistent phone partner for conversations across both sessions.

During each session, she posed questions in accordance with a pre-determined protocol. Although the scripts were subtly altered between sessions to foster engagement, they adhered to a consistent foundational structure. Importantly, the phone partner remained unaware whether her individual participants were interacting with a Hugvie or merely a mobile phone, ensuring an unbiased approach to the conversation.

The experimental procedures spanned two consecutive days: Days 1 and 2. To mitigate sequence effects, we implemented counterbalanced trial sessions between the two distinct groups (Figure 22.2). Participants in the

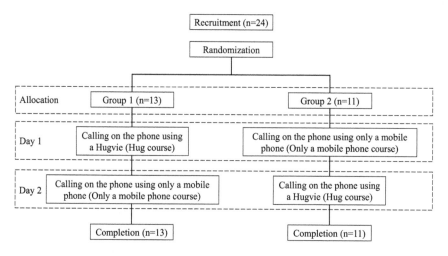

FIGURE 22.2 Experiment flowchart: Participants were first randomly assigned to two groups: Group 1 talked on phone using a Hugvie on Day 1 and used only a mobile phone on Day 2; Group 2 followed a reverse order of sessions [40].

first group (Group 1; $n=13$) engaged in Hugvie sessions on Day 1 and in Phone sessions on Day 2. Conversely, participants in Group 2 ($n=11$) followed a reverse session order. Each participant communicated with the same partner on both days. The average duration of these phone interactions was roughly ten minutes. To control for potential hormonal fluctuations due to diurnal rhythms, we scheduled each participant's phone conversations at identical times on both days and instructed them to abstain from eating one hour before them.

Following each session, participants completed questionnaires to gauge their self-confidence under both sessions. Responses were recorded on a Likert rating scale, with scores ranging from 0 (no comfort) to 6 (high comfort). Additionally, after completing both sessions, participants were asked a binary question: "Did you find it more comfortable to converse on the mobile phone while embracing Hugvie as opposed to using only the mobile phone?"

We assessed their physiological reactions by collecting saliva samples on Days 1 and 2 to measure their cortisol levels. Note that a time lag of approximately 20 minutes exists between an event's occurrence and the detection of related changes in salivary cortisol levels [29]. During the interaction, we collected from each participant the following four salivary cortisol samples: S1 (immediately prior to the conversation's commencement), S2 (20 minutes after its start), S3 (20 minutes after its end), and S4 (40 minutes after its end). These S1 measurements served as baseline values that reflect resting cortisol levels, and the S2, S3, and S4 measurements, respectively, corresponded to the cortisol levels at the beginning, end, and a period after the experience (Figure 22.3). Just prior to the conversation, each participant was taken to a private room where a research assistant

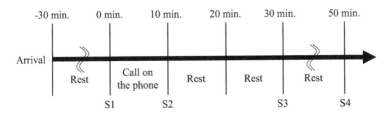

FIGURE 22.3 Timeline of events on Days 1 and 2. Both groups of participants (Hugvie users and only mobile phone users) were requested to relax for 30 minutes before and 40 minutes after talking on the phone. Salivary cortisol measures were collected just before the start of the phone conversation as a baseline measurement (S1), 20 minutes after the start of the conversation (S2), 20 minutes after the end of the conversation (S3), and 40 minutes after the end of the conversation [40].

collected a saliva sample. Participants were given a relaxation period in this room of 30 minutes before and 40 minutes after the phone conversations. They were only permitted to leave the room after providing the final saliva sample for the S4 measurement.

Saliva samples, ranging from 0.5 to 2.0 mL, were passively collected and promptly transferred to sterile, 15-mL plastic tubes. They were then flash-frozen on dry ice and preserved at −80°C until analysis time. Upon thawing at ambient temperature, the samples were centrifuged at $1,500 \times g$ for ten minutes at 4°C to eliminate large precipitates. The salivary cortisol levels were then quantified in duplicate using a cortisol enzyme immunoassay kit (Salimetrics, State College, PA, USA), and the sample treatments (25 μL) were done in accordance with the manufacturer's guidelines. Subsequently, the concentrations were computed using MATLAB 7, based on a corresponding standard curve [30].

22.2.3 Data Analysis

Statistical evaluations were conducted using SPSS version 24.0 (IBM, Armonk, NY, USA). We assessed the group variations through a *t*-test in the demographic data, including age, full-scale IQ, AQ-J score, LSAS score, ADHD-RS score, and AASP subscale scores. Gender distribution differences between groups were examined using a χ^2-test. We compared the self-confidence ratings and salivary cortisol levels to the baseline values (S2/S1, S3/S1, and S4/S1), adhering to the recommended statistical methodology for crossover trials, as previously described [31].

22.3 RESULTS

22.3.1 Demographic Data

We enrolled 24 ASD individuals in this study. Among them, one participant had an IQ below 70, 17 exhibited abnormally high scores on the AQ-J, and all 24 demonstrated social anxiety as indicated by their LSAS scores. None registered unusually high ADHD-RS scores. Within the group, six participants scored unusually high in low registration, two displayed sensation-seeking tendencies, five exhibited sensory sensitivity, and six showed sensory avoidance (Table 22.1). All the participants successfully completed both the experimental procedure and the associated questionnaires. When asked, "Did you find it more comfortable to converse on the mobile phone while embracing Hugvie as opposed to using only the mobile phone?" 21 participants (87.5%) expressed a preference for Hugvie. The remaining three stated, "I found neither option preferable."

TABLE 22.1   Descriptive Statistics of Participants in Group 1 and Group 2 (*n*=24)

Characteristics	Group 1 (*n* = 13) Mean (SD)	Group 2 (*n*=11) Mean (SD)	Statistics *p*
Age	20.3 (3.4)	20.2 (2.4)	0.92
Gender (male, female)	11, 2	10, 1	0.64
Full-scale IQ	88.4 (14.4)	86.9 (14.4)	0.81
AQ-J	31.2 (4.3)	34.1 (3.2)	0.08
LSAS	47.4 (8.7)	43.0 (10.0)	0.27
ADHD-RS	8.2 (3.5)	9.7 (5.4)	0.42
AASP			
Low registration	36.4 (7.4)	33.2 (9.3)	0.36
Sensation seeking	37.2 (10.8)	33.0 (5.7)	0.26
Sensory sensitivity	37.7 (9.6)	32.1 (11.5)	0.21
Sensation avoiding	37.5 (10.4)	37.0 (12.4)	0.91

AQ-J: Autism Spectrum Quotient-Japanese version.

LSAS: Liebowitz Social Anxiety Scale.

AASP: Adolescent/Adult Sensory Profile.

22.3.2 Main Result

Figure 22.4 shows the within-subject summation of the self-confidence ratings from Days 1 and 2 for each group. Statistical analysis revealed no significant carryover effects ($t(18.2)=-1.02$, $p=0.32$, $r=0.23$). We found a significant treatment effect between the groups ($t(19.9)=3.12$, $p=0.01$, $r=0.57$), showing an enhancement in self-confidence among participants in the Hugvie session compared to those in the Phone session.

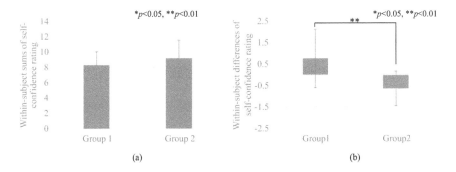

FIGURE 22.4   (a) Within-subject sum of the self-confidence ratings on Days 1 and 2; (b) Within-subject difference in the self-confidence ratings between Days 1 and 2 [40].

Figure 22.5 shows the within-subject summation of the changes in the salivary cortisol levels on Days 1 and 2 for each sampling point (i.e., S2, S3, and S4) within each group. There were no carryover effects in any of the sampling points (S2/S1: $t(21.4) = 1.34$, $p = 0.19$, $r = 0.28$; S3/S1: $t(18.4) = 1.17$, $p = 0.26$, $r = 0.26$; and S4/S1: $t(14.9) = 1.74$, $p = 0.10$, $r = 0.41$). We further evaluated the treatment effect for all the sampling points by considering the differences in the changes in the salivary cortisol levels between Days 1 and 2 for both groups. We found significant differences between participants in the Hugvie and Phone sessions across all the sampling points (S2/S1: $t(22.0) = -3.14$, $p = 0.05$, $r = 0.56$; S3/S1: $t(17.1) = -3.04$, $p = 0.01$, $r = 0.59$; and S4/S1: $t(16.9) = -3.82$, $p = 0.01$, $r = 0.68$). Figure 22.6 shows the within-subject differences in the changes in salivary cortisol

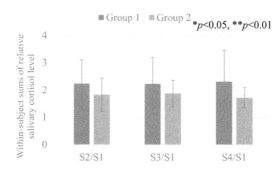

FIGURE 22.5 Within-subject sums of changes in salivary cortisol levels on Days 1 and 2 [40].

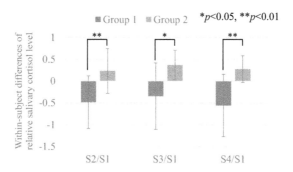

FIGURE 22.6 Within-subject differences in changes in salivary cortisol levels between Days 1 and 2: Group 1 participants called a stranger on phone while hugging a Hugvie on Day 1, followed by a phone call without it (only using a mobile phone) on Day 2. Group 2 participants followed opposite conditions on both days [40].

TABLE 22.2 Means and Standard Errors of Mean (SEM) of the Confidence Rating Scale after the Phone Call and Changes in Cortisol Levels in Hugvie Users and Only-Mobile-Phone Users in Group 1 and Group 2

	Day	
	Day 1	Day 2
Group	Mean (SEM)	Mean (SEM)
Self-confidence		
Group 1	4.54 (1.05)	3.77 (1.17)
Group 2	4.27 (0.38)	4.91 (0.37)
Change in cortisol level S2/S1		
Group 1	0.87 (0.10)	1.36 (0.18)
Group 2	1.03 (0.14)	0.80 (0.10)
Change in cortisol level S3/S1		
Group 1	0.94 (0.09)	1.29 (0.22)
Group 2	1.12 (0.10)	0.75 (0.08)
Change in cortisol level S4/S1		
Group 1	0.87 (0.11)	1.43 (0.24)
Group 2	1.00 (0.08)	0.72 (0.06)

levels between Days 1 and 2 for each sampling point within each group. Comprehensive details regarding the participants' confidence ratings after the conversations and the changes in the cortisol levels are provided in Table 22.2.

22.4 CONCLUSION AND DISCUSSION

In this section, we explored the potential advantages of employing a huggable pillow called Hugvie during phone conversations with unfamiliar individuals among young ASD adults. Our findings revealed that using Hugvie significantly mitigated their stress during the phone interactions, showing reduced cortisol levels in the participants' saliva across all the sampling instances. We also observed a marked enhancement in self-confidence among those who engaged with Hugvie in contrast to those who just relied on a mobile phone. These outcomes suggest that incorporating Hugvie into the conversations alleviated stress and enhanced their self-confidence. Consequently, our study underscores the potential benefits of using Hugvie in interactions with unfamiliar individuals, highlighting the significance of tactile stimulation for young adults with ASD in communicative contexts.

We objectively evaluated the self-confidence levels of our young adult participants by analyzing their salivary cortisol levels. The physiological arousal in response to social engagement is markedly elevated in ASD individuals compared to the general populace [32–34], and there is an approximate 20-minute time lag between an event's occurrence and the detection of correlated changes in salivary cortisol [29]. Our findings revealed significant changes in cortisol responses between those who interacted with a Hugvie and those who just used a mobile phone across all saliva-sampling intervals and measurements (S2/S1, S3/S1, and S4/S1). These outcomes suggest that embracing Hugvie mitigates stress following a phone conversation with an unfamiliar individual.

In touch interactions, the literature distinguishes between active and passive touch [35,36]. The latter entails the application of a tactile stimulus to the skin without a voluntary movement, while active touch involves an intentional movement to seek a tactile stimulus, thereby integrating tactile and proprioceptive information generated by the movement. Individuals tend to exhibit hypersensitivity and an inability to concentrate on the properties of the stimuli when they are touched passively [37]. Hugvie, on the other hand, encourages active touch. By enthusiastically engaging with it, ASD individuals can attune to its properties and achieve comfort. This observation leads us to posit that it is such active engagement with Hugvie, rather than its passive touch, that contributed to our outcomes in this study.

Note that ASD individuals often exhibit pronounced variability in their preferences for certain textures, as compared to those with typical development [38]. This observation aligns with the broader understanding that ASD individuals tend to have strong inclinations and aversions [39]. In the context of our study, we operated under the assumption that the texture of Hugvie would align with the preferences of our participants. We posit that Hugvie's potential as a supportive tool might be further enhanced and refined by pursuing an ideal texture that resonates with the unique sensory needs of ASD individuals.

ACKNOWLEDGMENT

This work was supported in part by Grants-in-Aid for Scientific Research from the Japan Society for the Promotion of Science (19H04880, 20H00101) and JST CREST Grant Number JPMJCR18A1 and JST and Moonshot R&D Grant Number JPMJMS2011.

REFERENCES

[1] Jaafar NI, Darmawan B, Mohamed MY. Face-to-face or not-to-face: a technology preference for communication. *Cyberpsychol Behav Soc Netw.* 2014; 17: 702–708. https://doi.org/10.1089/cyber.2014.0098 PMID: 25405782

[2] Vogely AJ. Listening comprehension anxiety: students' reported sources and solutions. *Foreign Lang Ann.* 1998; 31: 67–80.

[3] Gallace A, Spence C. The science of interpersonal touch: an overview. *Neurosci Biobehav Rev.* 2010; 34: 246–359. https://doi.org/10.1016/j.neubiorev.2008.10.004 PMID: 18992276

[4] Tatsukawa K, Nakano T, Ishiguro H, Yoshikawa Y. Eyeblink synchrony in multimodal human-android interaction. *Sci Rep.* 2016; 6: 39718. https://doi.org/10.1038/srep39718 PMID: 28009014

[5] Chang A, O'Modhrain S, Jacob R, Gunther E, Ishii H. ComTouch: design of a vibrotactile communication device. In: *Proceedings of Designing Interactive Systems.* New York: ACM Press;2002. pp. 312–320.

[6] Miguel OH, Sampaio A, Martínez-Regueiro R, Gómez-Guerrero L, López-Dóriga CG, Gómez S, et al. Touch processing and social behavior in ASD. *J Autism Dev Disord.* 2017; 47: 2425–2433. https://doi. org/10.1007/s10803-017-3163-8 PMID: 28534141

[7] Ballard DH, Hayhoe MM, Pook PK, Rao RP. Deictic codes for the embodiment of cognition. *Behav Brain Sci.* 1997; 20: 723–742. https://doi.org/10.1017/s0140525x97001611 PMID: 10097009

[8] Ybarra GJ, Passman RH, Eisenberg CS. The presence of security blankets or mothers (or both) affects distress during pediatric examinations. *J Consult Clin Psychol.* 2000; 68: 322–330. https://doi.org/10.1037//0022-006x.68.2.322 PMID: 10780133

[9] Sumioka H, Nakae A, Kanai R, Ishiguro H. Huggable communication medium decreases cortisol levels. *Sci Rep.* 2013; 3: 3034. https://doi.org/10.1038/srep03034 PMID: 24150186

[10] Yamazaki R, Christensen L, Skov K, Chang CC, Damholdt MF, Sumioka H, et al. Intimacy in phone conversations: anxiety reduction for Danish seniors with *Hugvie. Front Psychol.* 2016; 7: 537. https://doi. org/10.3389/fpsyg.2016.00537 PMID: 27148144

[11] Sumioka H, Nishio S, Minato T, Yamazaki R, Ishiguro H. Minimal human design approach for sonzaikan media: investigation of a feeling of human presence. *Cognit Comput.* 2014; 6: 760–774.

[12] Nakanishi J, Sumioka H, Ishiguro H. Impact of mediated intimate interaction on education: a huggable communication medium that encourages listening. *Front Psychol.* 2016; 7: 510. https://doi.org/10.3389/fpsyg.2016.00510 PMID: 27148119

[13] Foss-Feig JH, Heacock JL, Cascio CJ. Tactile responsiveness patterns and their association with core features in autism spectrum disorders. *Res Autism Spectr Disord.* 2012; 6: 337–344. https://doi.org/10.1016/j.rasd.2011.06.007 PMID: 22059092

[14] Cascio CJ, Lorenzi J, Baranek GT. Self-reported pleasantness ratings and examiner-coded defensiveness in response to touch in children with ASD: effects of stimulus material and bodily location. *J Autism Dev Disord*. 2016; 46: 1528–1537. https://doi.org/10.1007/s10803-013-1961-1 PMID: 24091471

[15] Rogers SJ, Hepburn S, Wehner E. Parent reports of sensory symptoms in toddlers with autism and those with other developmental disorders. *J Autism Dev Disord*. 2003; 33: 631–642. https://doi.org/10.1023/b:jadd.0000006000.38991.a7 PMID: 14714932

[16] Wodka EL, Puts NAJ, Mahone EM, Edden RAE, Tommerdahl M, Mostofsky SH. The role of attention in somatosensory processing: a multi-trait, multi-method analysis. *J Autism Dev Disord*. 2016; 46: 3232–3241. https://doi.org/10.1007/s10803-016.3866-6 PMID: 27448580

[17] Crane L, Goddard L, Pring L. Sensory processing in adults with autism spectrum disorders. *Autism*. 2009; 13: 215–228. https://doi.org/10.1177/136.3361309103794 PMID: 19369385

[18] Tavassoli T, Hoekstra RA, Baron-Cohen S. The Sensory Perception Quotient (SPQ): development and validation of a new sensory questionnaire for adults with and without autism. *Mol Autism*. 2014; 5: 29. https://doi.org/10.1186.3040-2392-5-29 PMID: 24791196

[19] Cascio C, McGlone F, Folger S, Tannan V, Baranek G, Pelphrey KA, et al. Tactile perception in adults with autism: a multidimensional psychophysical study. *J Autism Dev Disord*. 2008; 38: 127–137. https://doi.org/10.1007/s10803-007-0370-8 PMID: 17415630

[20] American Psychiatric Association. *Diagnostic and Statistical Manual of Mental Disorders*, 5th ed. Washington DC: American Psychiatric Association; 2013.

[21] Leekam SR, Libby SJ, Wing L, Gould J, Taylor C. The diagnostic interview for social and communication disorders: algorithms for ICD-10 childhood autism and Wing and Gould autistic spectrum disorder. *J Child Psychol Psychiatry*. 2002; 43: 327–342. https://doi.org/10.1111/1469-7610.00024 PMID: 11944875

[22] Wing L, Leekam SR, Libby SJ, Gould J, Larcombe M. The diagnostic interview for social and communication disorders: background, inter-rater reliability and clinical use. *J Child Psychol Psychiatry*. 2002; 43: 307–325. https://doi.org/10.1111/1469-7610.00023 PMID: 11944874

[23] Sheehan DV, Lecrubier Y, Sheehan KH, Amorim P, Janavs J, Weiller E, et al. The Mini-International Neuropsychiatric Interview (M.I.N.I.): the development and validation of a structured diagnostic psychiatric interview for DSM-IV and ICD-10. *J Clin Psychiatry*. 1998; 59(Suppl. 20): 22–33; quiz 4–57. PMID: 9881538

[24] Wakabayashi A, Tojo Y, Baron-Cohen S, Wheelwright S. The Autism-Spectrum Quotient (AQ) Japanese version: evidence from high-functioning clinical group and normal adults. *Shinrigaku Kenkyu*. 2004; 75: 78–84. https://doi.org/10.4992/jjpsy.75.78 PMID: 15724518

[25] Wechsler D. *Wechsler Intelligence Scale for Children-Fourth Edition (WISC-IV)*. San Antonio, TX: The Psychological Corporation; 2003.

[26] Liebowitz MR. Social phobia. Anxiety. *Mod Probl Pharmacopsychiatry*. 1987; 22: 141–173.

[27] DuPaul GJ, Reid R, Anastopoulos AD, Lambert MC, Watkins MW, Power TJ. Parent and teacher ratings of attention-deficit/hyperactivity disorder symptoms: factor structure and normative data. *Psychol Assess*. 2016; 38: 214–225. https://doi.org/10.1037/pas0000166 PMID: 26011476

[28] Brown C, Dunn W. *Adolescent-Adult Sensory Profile: User's* Manual. Therapy Skill Builders. Pearson: London, England. 2002.

[29] Kirschbaum C, Hellhammer DH. Salivary cortisol in psychobiological research: an overview. *Neuropsychobiology*. 1989; 22: 150–169. https://doi.org/10.1159/000118611 PMID: 2485862

[30] Tsuji S, Yuhi T, Furuhara K, Ohta S, Shimizu Y, Higashida H. Salivary oxytocin concentrations in seven boys with autism spectrum disorder received massage from their mothers: a pilot study. *Front Psychiatry*. 2015; 6: 58. https://doi.org/10.3389/fpsyt.2015.00058 PMID: 25954210

[31] Wellek S, Blettner M. On the proper use of the crossover design in clinical trials. Part 18 of a series on evaluation of scientific publications. *Dtsch Arztebl Int*. 2012; 109: 276–381. https://doi.org/10.3238/arztebl.2012.0276 PMID: 22567063

[32] Corbett BA, Schupp CW, Lanni KE. Comparing biobehavioral profiles across two social stress paradigms in children with and without autism spectrum disorders. *Mol Autism*. 2012; 3: 13. https://doi.org/10.1186.3040-2392-3-13 PMID: 23158965

[33] Schupp CW, Simon D, Corbett BA. Cortisol responsivity differences in children with autism spectrum disorders during free and cooperative play. *J Autism Dev Disord*. 2013; 43: 2405–2417. https://doi.org/10.1007/s10803-013-1790-2 PMID: 23430177

[34] Corbett BA, Gunther JR, Comins D, Price J, Ryan N, Simon D, et al. Brief report: theatre as therapy for children with autism spectrum disorder. *J Autism Dev Disord*. 2010; 41: 505–511.

[35] Ackerley R, Hassan E, Curran A, Wessberg J, Olausson H, McGlone F. An fMRI study on cortical responses during active self-touch and passive touch from others. *Front Behav Neurosci*. 2012; 6: 51. https://doi.org/10.3389/fnbeh.2012.00051 PMID: 22891054

[36] Olczak D, Sukumar V, Pruszynski JA. Edge orientation perception during active touch. *J Neurophysiol*. 2018; 120: 2423–2429. https://doi.org/10.1152/jn.00280.2018 PMID: 30133382

[37] Park M, Chung Y. An emotional comfort product using touch: the possibility. *Arch Des Res*. 2020; 33: 89–103.

[38] Cascio CJ, Moana-Filho EJ, Guest S, Nebel MB, Weisner J, Baranek GT, et al. Perceptual and neural response to affective tactile texture stimulation in adults with autism spectrum disorders. *Autism Res*. 2012; 5: 231–244. https://doi.org/10.1002/aur.1224 PMID: 22447729

[39] Trevarthen C, Delafield-Butt JT. Autism as a developmental disorder in intentional movement and affective engagement. *Front Integr Neurosci*. 2013; 7: 49. https://doi.org/10.3389/fnint.2013.00049 PMID: 23882192

[40] Sumioka H, Kumazaki H, Muramatsu T, Yoshikawa Y, Ishiguro H, Higashida H, et al. A huggable device can reduce the stress of calling an unfamiliar person on the phone for individuals with ASD. *PLoS One*. 2021; 16(7): e0254675. https://doi.org/10.1371/journal.pone.0254675

Viewing a Presenter's Touch Affects the Feeling of *Kawaii* of Others toward an Object

Yuka Okada, Mitsuhiko Kimoto, Takamasa Iio,
Katsunori Shimohara, Hiroshi Nittono,
and Masahiro Shiomi

Advanced Telecommunications Research
Institute International, Kyoto, Japan

Doshisha University, Kyoto, Japan

Osaka University, Osaka, Japan

23.1 INTRODUCTION

A feeling of cuteness is one essential factor in promoting people's positive emotions and their behavioral changes [1–5]. In the context of human–robot interaction, the importance of the concept of cuteness has been broadly investigated [6,7]. In human–robot interaction studies, the concept of *kawaii* (a Japanese word that means "cute" [5,8]) is typically addressed in two distinct ways: (1) accentuating the perceived feelings

DOI: 10.1201/9781003384274-29

of *kawaii* through a robot's appearance and (2) expressing this feeling to others by actions. The design and appearance of these commercial robots reflect the positive connotations of *kawaii*, a critical element in Japan's commercial and pop culture landscape [9,10]. While the first approach has been widely employed in consumer robotics, including Paro, LOVOT, and Robohon, the second approach has been less thoroughly explored.

The baby schema [11,12], a well-known concept in *kawaii* or cute contexts, has also been adopted in existing products and social robot designing in academia [13]. However, its focus is predominantly on appearance rather than behavior. In contrast, research remains limited on designing behaviors that convey a sense of *kawaii*. A previous study investigated potential locomotion behaviors that express *kawaii* feelings with a mobile robot called a Roomba [14].

Like these previous studies that investigated the design of robot behaviors to convey the feeling of *kawaii*, we are also interested in the behavior design of robots for increasing and conveying feelings of *kawaii*. To achieve such behavior, we focus on a possible role for social robots in real environments. We believe that social robots, like sales clerks in a shopping mall, will recommend products in the near future by emphasizing the taste of food, the utility of gadgets, the affordability of items, and the cuteness of dolls [15–17]. In such a context, to foster natural and smooth interactions with people, affective, social robots capable of emotional expressions must be designed [18–20]. As an example, a study examined the relationship between people's attributions of different emotions to robots and the latter's anthropomorphism [21]. Expressing *kawaii* feelings, along with other affective and emotional behaviors, positively influenced the perceived anthropomorphism of robots and the impressions they made on people [22,23]. This suggests that expressing *kawaii* feelings is an essential function for social robots that interact with humans in actual typical environments.

Focusing on the previously proposed social effects concept of a "*kawaii* triangle" [8], we aim to highlight the *kawaii* feelings toward objects through robots. This concept is illustrated when person X sees person Y's smile in response to her *kawaii* feeling toward an item (e.g., a penguin), and she then develops a positive impression of both person Y and the penguin. Person X's expression of *kawaii* feelings can also enhance person

Y's *kawaii* feelings. With this in mind, we hypothesized that an observer's *kawaii* feelings toward both a social robot and a specific item will be heightened if the social robot displays a stronger expression of *kawaii* feelings toward that specific item.

In this study, we explored the impact of a social robot's touching and exaggerated behaviors on an observer's perception of *kawaii*, because a touching behavior is a typical approaching behavior caused by a feeling of *kawaii* [8]. Based on the kawaii-triangle concept, we hypothesized that an observer will perceive that a *kawaii* feeling from a robot toward an object is strengthened when the robot touched it (Figure 23.1), i.e., approaching it, leading to increased *kawaii* feelings in the observer for both the object and the robot. Moreover, previous studies identified "cute aggression" as a relationship between aggressive behaviors and strong *kawaii* feelings [24,25]. Consequently, we also hypothesize that participants will experience stronger *kawaii* feelings toward both the robot and the object if a social robot exhibits exaggerated touching behavior.

To better understand the impact of social robot behavior on an observer's perception of *kawaii* feelings and strengthen the design of engaging and

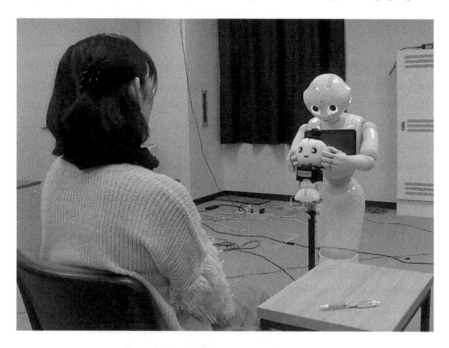

FIGURE 23.1   Robot explains a doll with a touching behavior.

emotionally expressive robots, we conducted Experiment I with a robot named Pepper and a Pepper doll as objects and two explanatory behaviors (touch factors: touch and no-touch) and two action types (motion factors: normal and emphasis) to address the following research questions:

Research question 1: Does observing a robot's touching behavior heighten the perceived *kawaii* feelings of the robot toward the object, and does this observation also strengthen the observer's *kawaii* feelings for the robot and the object?

Research question 2: Does exaggerated explanatory behavior enhance the observer's perception of the robot's *kawaii* feelings toward the object and elicit a stronger *kawaii* response from the observer to both the object and the robot?

Note that this chapter is modified based on our previous work [26], edited to be comprehensive and fit with the context of this book.

23.2 MATERIALS AND METHODS OF EXPERIMENT

We explored the potential impact of observing a robot's touching behaviors and its demonstration of *cute aggression* on the observer's perception of the robot's feelings of *kawaii* toward an object. We also examined the influence of these behaviors on the observer's own *kawaii* feelings toward the robot and an object. This study received approval from the ATR Review Board Ethics Committee (20–501–4).

23.2.1 Robot, Task, and Environment

We chose Pepper for our experiment. Standing 121 cm tall, Pepper has 20 degrees of freedom (DOFs), including two in its head, six in each arm, and six in its lower body. It interacted with a 28-cm tall doll that resembled Pepper itself. We positioned the doll between Pepper and the participant and placed it on a 65 cm stand to equalize their heights.

Pepper introduced itself and explained the doll's four characteristics: costume, tactile sensation, shape, and facial design. The robot's speech for each characteristic is as follow: "Unlike me, this Pepper wears a tuxedo. Very stylish." (Costume), "It feels fluffier and cuddlier than me." (Tactile sensation), "Although I don't really think I'm fat, this doll does resemble me." (Shape), and "I like its big face and round eyes." (Facial design). Through this setup, we assessed how its touching behaviors and its manifestation of a *cute aggression* design impacted the observer's perception of *kawaii*.

23.2.2 Conditions

We adopted a within-subjects experimental design. Each participant was engaged in four trials, encompassing two touch conditions (*touch* and *no-touch*) and two motion conditions (*normal* and *emphasis*). To mitigate any potential impact of the order effects, we counterbalanced the sequences of these experimental conditions.

23.2.2.1 Touch Factor

In the *touch* condition, we illustrated different aspects of the doll described above by developing four specific actions: touching the doll's body (to convey costume details), stroking its head (to express tactile sensation), touching its feet (to indicate shape), and squeezing its cheeks (to illustrate facial design).

For the no-touch condition, we designed four alternate actions, which involved the robot spreading its hands around the doll's body, head, and feet without physical contact. We based this design on previous studies that highlighted the effectiveness of deictic and iconic gestures in tasks involving information provision [27–30]. These hand motions depicted the doll's shape and drew the observer's attention to the parts being described.

23.2.2.2 Motion Factor

In designing the motion factor, we had two conditions: *normal* and *emphasized*. To settle upon an appropriate speed for the *normal* condition, we asked participants in a preliminary study to interact with the doll and freely express their feelings of *kawaii*. Based on their actions, which included behaviors like stroking, squeezing, and touching, we adjusted the robot's speed to mimic these human behaviors. We, however, had to make heuristic adjustments due to the physical differences between humans and robots.

For the *emphasized* condition, we chose a motion speed three times faster than normal speed, based on heuristic adjustments and multiple discussions among the authors. Despite previous studies indicating the effectiveness of enhancing pitch for information delivery tasks [29,30], we concentrated on the robot's gestures, specifically its touch behavior and style, when expressing feelings of *kawaii*. This decision was guided by our focus on investigating the concept of *cute aggression* rather than the combined effects of gestures and speech.

23.2.3 Measurement

Our study employed a two-item questionnaire to measure the feelings of *kawaii*: (1) the degree of *kawaii* and (2) the degree of wanting to approach.

We chose these specific items based on an existing research [5,8] that reported a positive correlation between feelings of *kawaii* and a motivation to approach. These items were measured across three targets: the robot's feelings toward the doll, the participant's feelings toward the doll, and the participant's feelings toward the robot.

In addition to the feelings of *kawaii*, we also used a questionnaire to evaluate the impressions of the robot's presentation: the quality of the presentation and the naturalness of the robot's motions. All responses were gathered using a seven-point response format, where 1 signified a negative response and 7 indicated a positive response. Participants were also given an open-ended response form for additional feedback.

23.2.4 Procedure

After our participants gave written consent, we led them through the experiment's procedure and requested that they envision a scenario in which a shopkeeper was recommending a doll. In each session, the robot performed a standardized sequence of actions, where only the touch behaviors and their motion styles differed between conditions. In each session, the robot initiated the interaction with a greeting, described the doll, and concluded the dialogue. After every session, participants filled out a questionnaire, providing feedback and impressions.

23.2.5 Participants

We recruited a diverse group of 42 participants for the experiment, with an even gender split and ages ranging from 21 to 49 (average age: 37.83 years, *SD*: 7.92).

23.3 RESULT AND DISCUSSION

23.3.1 Questionnaire Results

Figures 23.2–23.5 display the results of a two-way repeated ANOVA, including averages and standard errors (*SE*) for the perceived feelings of *kawaii* and *wanting to approach*, the robot's feeling toward the doll, the participant's feeling toward the robot, and the participant's feeling toward the doll. Given the large number of combinations, this section only discusses the parts where significant differences emerged in the analysis.

The analysis results of the perceived robot's feelings of *kawaii* and *wanting to approach* the doll only showed significant differences in the touch factor (Figure 23.2). The analysis results of the perceived participant's feelings of *kawaii* and *wanting to approach* the robot did not show significant differences in any of the factors (Figure 23.3). The analysis results of the perceived participant's feelings of *kawaii* and *wanting to approach* the doll showed significant differences in all of the factors (Figure 23.4). For the *kawaii* feelings, the simple main effects showed significant differences: *touch> no-touch, p* < 0.001 in the *normal* condition, and *emphasized> normal, p*<0.001 in the *no-touch* condition. For *wanting to approach,* the simple main effects showed significant differences: *touch> no-touch, p* < 0.001 in the *normal* condition, and *emphasized> normal, p*<0.001 in

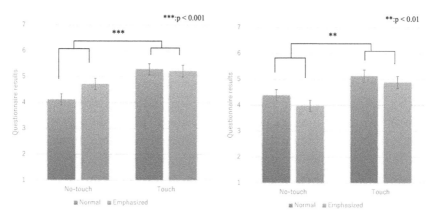

FIGURE 23.2   Perceived robot's feeling of *kawaii* and wanting to approach the doll.

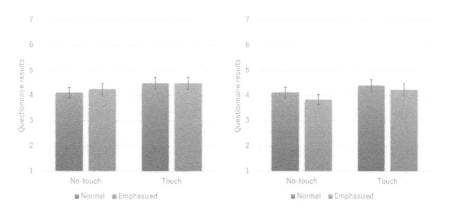

FIGURE 23.3   Perceived participant's feeling of *kawaii* and wanting to approach the robot.

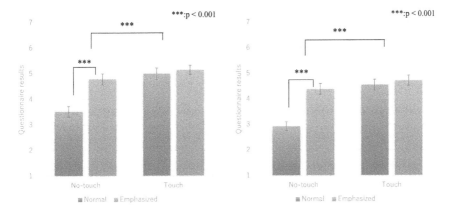

FIGURE 23.4 Perceived participant's feeling of *kawaii* and wanting to approach the doll.

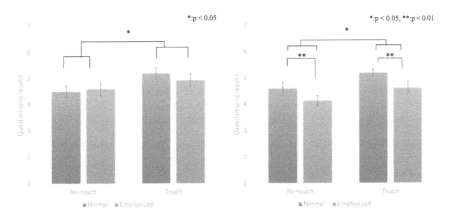

FIGURE 23.5 Impressions of robot's presentation and naturalness toward motions.

the *no-touch* condition. The analysis results of the participants' feelings about a *good description* and the *naturalness of all the motions* showed significant differences in the touch factor (Figure 23.5).

23.3.2 Summary of Questionnaire Analysis

Our questionnaire analysis summary yielded several insights. The robot's touching behavior elicited stronger feelings of *kawaii* and a higher degree of *wanting to approach* the doll compared to the non-touching behavior. It also enhanced the robot's impression toward the doll and outperformed the normal non-touching style. Moreover, the touch behavior resulted in a more favorable evaluation of the robot's explanation. However, the robot's touch behavior did not amplify the participants' feeling of *kawaii* or their degree of *wanting to approach* the robot compared to the robot's no-touch behavior in the normal style. This partially supports research question 1; while the touching behavior enhanced the robot's and the observers' *kawaii* feelings toward the object, it did not affect the observer's *kawaii* feelings toward the robot.

Additionally, the emphasis style showed no notable effect on the touch behavior. Therefore, the findings did not support research question 2; exaggerated explanatory behaviors did not intensify the robot's *kawaii* feelings toward the object. However, the emphasis style did yield results in the no-touch condition. Compared to the normal style, the robot's no-touch behavior in the emphasized style increased the kawaii feeling and the degree of *wanting to approach* the doll and the participants' *kawaii* feelings and motivation to the degree of *wanting to approach* the doll.

23.3.3 Limitations

Our study is limited in the following ways. It exclusively used a specific robot (Pepper), focused on a single emotion (*kawaii*), and employed heuristic parameters for designing behaviors. Ironically, these limitations also open avenues for future research. Different robots, a variety of emotions, and unique parameters can be explored in the future to better understand the effects of touch behavior. For example, using quite human-like appearance robots [31,32], different characterized robots [33,34], and machine-like robots [35] would enable us to investigate the effects of appearances in the context of the perceived feelings of *kawaii* in information-providing tasks. In addition, the hardware limitations of the robot added another layer of complexity as it was challenging to design strong aggressive behaviors akin to those observed in humans. Past studies showed that touch characteristics are essential to express different emotions by touch interaction [36–39]; therefore, showing different kinds of touch behaviors would have various effects on perceived impressions.

Another possible future work is to investigate the effects of direct touch interaction between humans and robots on the perceived feelings of *kawaii*. Past studies reported that active touch interaction from robots to people showed several behavior-change effects and provided positive effects [40–42]. Since touch interaction from people to robots also has positive effects on stress buffering [43], direct touch interaction may have influenced the perceived feelings of *kawaii*. Our work sets the stage for future investigations into *kawaii* behavior designs. Researchers aiming to create *kawaii* expressions can use our insights as a fundamental starting point.

23.4 CONCLUSION

We investigated the impact of a presenter's touch behavior on an object and the associated motion styles for enhancing the perceived feeling of *kawaii* and the overall impression of a presentation in an information-providing context. Our results revealed that when participants observed a presenter's touch behaviors, their perceived feelings of *kawaii* and the degree of *wanting to approach* the object increased compared to when no touch behaviors were present. Additionally, the presenter's perceived impression of the object improved significantly when touch behaviors were employed, as opposed to the normal non-touching style without any notable consequences for the presenters themselves. These results demonstrate the effectiveness of incorporating touch behaviors in conveying and amplifying the perceived feelings of *kawaii* for specific objects, a strategy that can be applied to robot presenters.

ACKNOWLEDGMENT

This research work was supported in part by JST CREST Grant Number JPMJCR18A1, Japan, and JSPS KAKENHI JP21H04897.

REFERENCES

[1] J. G. Myrick, "Emotion regulation, procrastination, and watching cat videos online: Who watches Internet cats, why, and to what effect?," *Computers in Human Behavior*, vol. 52, pp. 168–176, 2015.

[2] A. M. Proverbio, V. D. Gabriele, M. Manfredi, and R. Adorni, "No race effect (ORE) in the automatic orienting toward baby faces: When ethnic group does not matter," *Psychology*, vol. 2, no. 9, pp. 931–935, 2011.

[3] H. Nittono, and N. Ihara, "Psychophysiological responses to kawaii pictures with or without baby schema," *SAGE Open*, vol. 7, no. 2, p. 2158244017709321, 2017.

[4] K. Nishiyama, K. Oishi, and A. Saito, "Passersby Attracted by Infants and Mothers' Acceptance of Their Approaches: A Proximate Factor for Human Cooperative Breeding," *Evolutionary Psychology*, vol. 13, no. 2, p. 147470491501300210, 2015.

[5] H. Nittono, M. Fukushima, A. Yano, and H. Moriya, "The power of kawaii: Viewing cute images promotes a careful behavior and narrows attentional focus," *PloS One*, vol. 7, no. 9, p. e46362, 2012.

[6] C. Caudwell, C. Lacey, and E. B. Sandoval, "The (Ir)relevance of robot cuteness: An exploratory study of emotionally durable robot design," In *Proceedings of the 31st Australian Conference on Human-Computer-Interaction*, Fremantle, WA, Australia. pp. 64–72, 2019.

[7] C. Caudwell, and C. Lacey, "What do home robots want? The ambivalent power of cuteness in robotic relationships," *Convergence*, p. 1354856519837792, 2019.

[8] H. Nittono, "The two-layer model of 'kawaii': A behavioural science framework for understanding kawaii and cuteness," *East Asian Journal of Popular Culture*, vol. 2, no. 1, pp. 79–96, 2016.

[9] S. Kinsella, "Cuties in Japan," In *Women, media and consumption in Japan*, pp. 230–264: Routledge, 2013.

[10] S. Lieber-Milo, and H. Nittono, "From a word to a commercial power: A brief introduction to the Kawaii aesthetic in contemporary Japan," *Innovative Research in Japanese Studies*, vol. 3, pp. 13–32, 2019.

[11] K. Lorenz, "The innate forms of potential experience," *Zeitschrift für Tierpsychologie*, vol. 5, pp. 235–409, 1943.

[12] V. Brooks, and J. Hochberg, "A psychophysical study of "cuteness"," *Perceptual and Motor Skills*, vol. 11, p. 205, 1960.

[13] C.-H. Chen, and X. Jia, "Research on the influence of the baby schema effect on the cuteness and trustworthiness of social robot faces," *International Journal of Advanced Robotic Systems*, vol. 20, no. 3, p. 17298806231168486, 2023.

[14] M. Ohkura, "Kawaii engineering," In *Kawaii engineering*, pp. 3–12: Springer, 2019.

[15] M. Niemelä, P. Heikkilä, H. Lammi, and V. Oksman, "A social robot in a shopping mall: Studies on acceptance and stakeholder expectations," In *Social Robots: Technological, Societal and Ethical Aspects of Human-Robot Interaction*, pp. 119–144: Springer, 2019.

[16] T. Iio, S. Satake, T. Kanda, K. Hayashi, F. Ferreri, and N. Hagita, "Human-like guide robot that proactively explains exhibits," *International Journal of Social Robotics*, vol. 12, pp. 549–566, 2019.

[17] P. Heikkilä, H. Lammi, M. Niemelä, K. Belhassein, G. Sarthou, A. Tammela, A. Clodic, and R. Alami, "Should a robot guide like a human? A qualitative four-phase study of a shopping mall robot," In *International Conference on Social Robotics*, Madrid, Spain. pp. 548–557, 2019.

[18] B. R. Duffy, "Anthropomorphism and the social robot," *Robotics and Autonomous Systems*, vol. 42, no. 3–4, pp. 177–190, 2003.

[19] R. Kirby, J. Forlizzi, and R. Simmons, "Affective social robots," *Robotics and Autonomous Systems*, vol. 58, no. 3, pp. 322–332, 2010.

[20] M. F. Jung, "Affective grounding in human-robot interaction," In *2017 12th ACM/IEEE International Conference on Human-Robot Interaction (HRI)*, Vienna, Austria. pp. 263–273, 2017.

[21] N. Spatola, and O. A. Wudarczyk, "Ascribing emotions to robots: Explicit and implicit attribution of emotions and perceived robot anthropomorphism," *Computers in Human Behavior*, vol. 124, p. 106934, 2021.

[22] N. Spatola, S. Monceau, and L. Ferrand, "Cognitive impact of social robots: How anthropomorphism boosts performances," *IEEE Robotics & Automation Magazine*, vol. 27, no. 3, pp. 73–83, 2019.

[23] T. Law, J. de Leeuw, and J. H. Long, "How movements of a non-humanoid robot affect emotional perceptions and trust," *International Journal of Social Robotics*, vol. 13, pp. 1–12, 2020.

[24] O. R. Aragón, M. S. Clark, R. L. Dyer, and J. A. Bargh, "Dimorphous expressions of positive emotion: Displays of both care and aggression in response to cute stimuli," *Psychological Science*, vol. 26, no. 3, pp. 259–273, 2015.

[25] K. K. Stavropoulos, and L. A. Alba, ""It's so cute I could crush it!":
Understanding neural mechanisms of cute aggression," *Journal of Frontiers in Behavioral Neuroscience*, vol. 12, p. 300, 2018.

[26] Y. Okada, M. Kimoto, T. Iio, K. Shimohara, H. Nittono, and M. Shiomi, "Kawaii emotions in presentations: Viewing a physical touch affects perception of affiliative feelings of others toward an object," *PLoS One*, vol. 17, no. 3, p. e0264736, 2022.

[27] P. Bremner, and U. Leonards, "Speech and gesture emphasis effects for robotic and human communicators—A direct comparison," In *2015 10th ACM/IEEE International Conference on Human-Robot Interaction (HRI)*, Portland, United States. pp. 255–262, 2015.

[28] P. Bremner, and U. Leonards, "Iconic gestures for robot avatars, recognition and integration with speech," *Frontiers in Psychology*, vol. 7, p. 183, 2016.

[29] T. Bickmore, L. Pfeifer, and L. Yin, "The role of gesture in document explanation by embodied conversational agents," *International Journal of Semantic Computing*, vol. 2, no. 1, pp. 47–70, 2008.

[30] T. W. Bickmore, L. M. Pfeifer, and B. W. Jack, "Taking the time to care: Empowering low health literacy hospital patients with virtual nurse agents," In *Proceedings of the SIGCHI Conference on Human Factors in Computing Systems*, Boston, Massachusetts, United States. pp. 1265–1274, 2009.

[31] D. F. Glas, T. Minato, C. T. Ishi, T. Kawahara, and H. Ishiguro, "Erica: The erato intelligent conversational android," In *2016 25th IEEE International Symposium on Robot and Human Interactive Communication (RO-MAN)*, pp. 22–29, 2016.

[32] M. Shiomi, H. Sumioka, K. Sakai, T. Funayama, and T. Minato, "SŌTO: An android platform with a masculine appearance for social touch interaction," In *Companion of the 2020 ACM/IEEE International Conference on Human-Robot Interaction*, Cambridge, United Kingdom, pp. 447–449, 2020.

[33] R. Matsumura, M. Shiomi, K. Nakagawa, K. Shinozawa, and T. Miyashita, "A desktop-sized communication robot: "Robovie-mr2"," *Journal of Robotics and Mechatronics*, vol. 28, no. 1, pp. 107–108, 2016.

[34] R. Matsumura, and M. Shiomi, "An animation character robot that increases sales," *Applied Sciences*, vol. 12, no. 3, p. 1724, 2022.

[35] T. Kanda, H. Ishiguro, T. Ono, M. Imai, and R. Nakatsu, "Development and evaluation of an interactive humanoid robot "Robovie"," In *Proceedings 2002 IEEE International Conference on Robotics and Automation (Cat. No.02CH37292)*, Washington, DC, United States. vol. 2, pp. 1848–1855, 2002.

[36] X. Zheng, M. Shiomi, T. Minato, and H. Ishiguro, "What kinds of robot's touch will match expressed emotions?," *IEEE Robotics and Automation Letters*, vol. 5, pp. 127–134, 2019.

[37] X. Zheng, M. Shiomi, T. Minato, and H. Ishiguro, "How can robot make people feel intimacy through touch?," *Journal of Robotics and Mechatronics*, vol. 32, no. 1, pp. 51–58, 2020.

[38] X. Zheng, M. Shiomi, T. Minato, and H. Ishiguro, "Modeling the timing and duration of grip behavior to express emotions for a social robot," *IEEE Robotics and Automation Letters*, vol. 6, no. 1, pp. 159–166, 2020.

[39] M. Shiomi, X. Zheng, T. Minato, and H. Ishiguro, "Implementation and evaluation of a grip behavior model to express emotions for an android robot," *Frontiers in Robotics and AI*, vol. 8, p. 755150, 2021.

[40] K. Nakagawa, R. Matsumura, and M. Shiomi, "Effect of robot's play-biting in non-verbal communication," *Journal of Robotics and Mechatronics*, vol. 32, no. 1, pp. 86–96, 2020.

[41] M. Shiomi, and N. Hagita, "Audio-visual stimuli change not only robot's hug impressions but also its stress-buffering effects," *International Journal of Social Robotics*, vol. 13, pp. 469–476, 2021.

[42] M. Shiomi, A. Nakata, M. Kanbara, and N. Hagita, "Robot reciprocation of hugs increases both interacting times and self-disclosures," *International Journal of Social Robotics*, vol. 13, pp. 353–361, 2021.

[43] N. Geva, F. Uzefovsky, and S. Levy-Tzedek, "Touching the social robot PARO reduces pain perception and salivary oxytocin levels," *Scientific Reports*, vol. 10, no. 1, p. 9814, 2020.

Index

Note: **Bold** page numbers refer to tables and *italic* page numbers refer to figures.

9 781032 470269